A Beginner's Guide to the Universe

A Beginner's Guide to the Universe is a fascinating introduction to astronomy and the wonders of the night sky. It begins by looking at the universe as a whole, and describes what we can see in the night sky. The solar system is then explored in detail, taking each planet in turn, from the hot world of Mercury near the Sun, to the distant, frozen world of Pluto. Moons, asteroids, meteoroids and comets are described, and objects outside our solar system are explained. You will learn what stars are, and how they cluster together to form galaxies that allow us to map out the furthest reaches of our Universe. At the end of the book, Professor John Brown, Astronomer Royal for Scotland, answers astronomy questions posed by children. Although this book is written for children, it will also appeal to adults who wish to learn about astronomy for the first time.

Dr. Andrew Conway is a lecturer in astronomy, based in the Department of Physics and Astronomy at the Open University, UK. He is an active researcher in solar physics, and has given many lecture courses on popular astronomy. He is a regular guest lecturer at astronomical societies.

Rosie Coleman, mother of three, is a primary school teacher in Scotland, UK, and has taught children for almost thirty years. She has given many educational talks, and produced material for use in schools throughout Scotland.

A BEGINNER'S GUIDE
to the
UNIVERSE

Dr Andrew Conway
and **Rosie Coleman**

CAMBRIDGE
UNIVERSITY PRESS

PUBLISHED BY THE PRESS SYNDICATE OF THE UNIVERSITY OF CAMBRIDGE
The Pitt Building, Trumpington Street, Cambridge, United Kingdom

CAMBRIDGE UNIVERSITY PRESS
The Edinburgh Building, Cambridge CB2 2RU, UK
40 West 20th Street, New York, NY 1011–4211, USA
477 Williamstown Road, Port Melbourne, VIC 3207, Australia
Ruiz de Alarcón 13,28014 Madrid, Spain
Dock House,The Waterfront, Cape Town 8001, South Africa

http://www.cambridge.org

First published 2003

Printed in the United Kingdom at the University Press, Cambridge

Typeface Swift Regular 11/14.5 pt   *System* QuarkXPress® [SE]

A catalogue record for this book is available from the British Library

Library of Congress Cataloguing in Publication data

ISBN 0 521 80693 3 hardback

Contents

Our View from the Earth   1
Light-years   4
Looking back in time   5
The changing sky   6
Constellations   10
Touring the stars   14

The Solar System   46
How do we know about the Solar System?   46
How did people work out that the Earth is a sphere?   47
The night sky   49
Gravity   50
A star is born   51
The Sun, our star   54
The nine planets   62
Smaller things   91
How long will the Solar System last?   98

Stars and Galaxies   99
Dots of light   99
Look, but you can't touch   99
The big clue: star colours   101
Snapshots of the stars   103

Contents

The life of stars   106
The death of stars   108
Dead stars   111
Nebulae: clouds in space   114
Star clusters   120
Galaxies   121
The Universe   126
Life elsewhere in the Universe   128

Questions and Answers   130

How long does it take to go to a star?   132
When is the Sun going to blow up?   133
What is a super-cluster?   133
What triggers the nuclear explosions when a star is born? 134
How do you become an astronaut?   134
Do people think there is a tenth planet?   135
Will there ever be any kind of animals from Earth going
  into space?   135
Is there an end to the Universe or does it go on forever? 135
How long can you stay in space for?   136
Is it true that from other galaxies our Sun will look like a star? 136
How many stellar systems are there?   137
How did the Solar System form?   138
How did the dust form to make the planets?   138
If the Sun is a star, how does it come out in the daytime and
  other stars come out at night?   138
How do you know there's a black hole?   139
Can you explain the activity inside a black hole?   139
Have we got any evidence that there are aliens on Mars?   140
Is the Moon burning gas like the Sun and stars?   140
How big is a crater on the Moon?   140
How far is the Moon from Earth?   141
Will there be football stadiums on the Moon?   141
How long does it take to get to the Moon?   141
How could you take photographs of Venus when it is so hot? 142
Why did you go to space in the first place?   142

Is it true that if we went close enough to Jupiter it would suck
    us in like a vacuum cleaner? 142
Is it true that aliens landed in Roswell? 142
What stops neutron stars from getting any smaller than they
    end up and becoming black holes? 143
What can't astronomers explain? 144

Our View from the Earth

This chapter is all about our view of the Universe from the Earth and how this view changes as the Earth and things near the Earth move around.

Maybe you have stood on top of a hill, looked from the top of a tall building, or even looked out of a plane as it is taking off or landing. If you have then you know that you can see a spectacular view. You can see things that are near and things that are far away at the same time. You will have noticed that as things move further away they seem to become smaller. Maybe you have been so high that when you look down on people they look as small as ants. If you look through binoculars or a telescope then you can see much more. You might be able to read the writing on far away signs or see people's faces or watch what animals are doing without them seeing you.

When you look at the night sky you are also seeing a view. Once again you can see many things – some nearer and some further away. Things called **comets**, **meteors**, **asteroids**, the **Sun**, the **Moon** and the **planets** are all quite near to us. They are all in what we call the **Solar System**. Things called **stars**, **nebulae** and **galaxies** are much further away. The view you see on a clear night is what the **Universe** looks like from the Earth.

Perhaps you find it strange to think of anything in the night sky being 'near' to us. Words like 'near' and 'far' can mean different things depending on where you are and what you are doing. If you were at a zoo, then you might say you are 'near' an animal if you could reach out and touch it through the bars of its cage. The word 'near' means an arm's length away. If you were telling someone how to get to your local shop,

All the objects in this photograph are at very different distances from the person who took it. The trees are closest, then the Moon, then Mars (by the trees, on the right), then Jupiter (well above and slightly to the right of the Moon) and then Saturn (at the same height as the Moon to the right).

Image opposite. The constellation of Orion.

you might call it 'near' if it was a five-minute walk away. Now the word 'near' means much longer than an arm's length away. Words like 'near', 'far', 'small', 'big', 'hot' and 'cold' all mean different things to different people at different times.

In **Astronomy**, all of these words are used, although the distances and sizes are so huge that you can't easily imagine them. We say that the Moon is near the Earth because it is much closer to the Earth than anything else in the night sky, but the distance to the Moon is unimaginably larger than the distance to your local shop!

Because the stars are all so far away they don't seem to move at all. During your whole life you will always see the same star patterns.

The proper name for a pattern of stars is a **constellation**. The photo shows Orion, just one of the constellations. When your grandparents were your age they would have seen exactly the same constellations that you see today. Perhaps one day you will have children and grandchildren. They will all see these same constellations too.

Light-years

When we are talking about distances to stars, we need to know a little about light and how fast it travels. Light is the fastest thing in the Universe. It travels 300 000 kilometres every second. A very fast car would take a month to cover that distance. The Sun is 150 million kilometres from the Earth. It takes just over eight minutes for the Sun's light to reach your eyes. This means you are seeing the Sun as it was eight minutes ago. If the Sun were to blow up – which it won't do – then we wouldn't even know about it until eight minutes after it happened.

All the other stars are very far away from us. They are so far away that astronomers don't even try to talk about the distances in miles or kilo-metres. Instead, astronomers measure the distances in **light-years**. One light-year is the distance that light travels in one year. It is about 10 000 billion kilometres – far too big a number to imagine. In just four seconds light travels more than a million kilometres!

As well as light-years, you can talk about light-seconds, light-minutes, light-hours and light-days. A light-second is the distance that light travels in a second, a light-minute is the distance that light travels in a minute and so on.

It is important to remember that nothing travels faster than light. Let's imagine we have a spaceship that can travel at the speed of light. Let's also imagine that you set off in it to travel to **Alpha Centauri**, the nearest star to our Sun, starting at 9 o'clock in the morning on your best friend's 10th birthday.

Time
Hours.Minutes.Seconds

9.00.00   *Your friend waves goodbye to you.*
You set off.

9.00.01   *Your friend blinks once.*
You are at the Moon.

9.08.33   *Your friend sheds a few tears for you.*
You are further away than the Sun.

> 10.00.00   *Your friend has spent an hour at school.*
> *You are between Jupiter and Saturn.*
>
> *By 9 o'clock the next morning, when your friend is having breakfast,*
> *you have passed Pluto, the furthest planet from the Sun.*
> *By your friend's 14th birthday you are almost at the nearest star.*

Looking back in time

The star Alpha Centauri is over four light-years away from us, so the light from this star takes over four years to get here. This means that if you look at Alpha Centauri you are seeing light that left the star over four years ago. If a star is ten light-years away, you see it as it looked ten years ago and if a star is 100 light-years away, you see it as it looked 100 years ago. Some of the stars we can see are so far away that their light has taken millions of years to reach us. This means that as you gaze at these stars you are looking back in time.

How far can you see?

On a clear night, without a telescope or binoculars, you can see a fuzzy patch in the constellation Andromeda. It is The Andromeda Galaxy. It is over two million light-years away from the Earth. You are seeing light which left the Andromeda Galaxy two million years ago. It is the most distant object our eyes can see without a telescope.

Dazzle your friends with your brilliance!

Ask them how far the human eye can see without binoculars or a telescope and watch their surprise as you tell them the answer.

5

The changing sky

To many people who lived on the Earth long ago the night sky was like a stained-glass window or a beautiful picture. It showed them freeze-framed pictures of stories about their gods, heroes, sacred objects and animals. Although the pictures themselves don't change much over the years, the sky is always changing and always moving.

Imagine, like ancient people must have done, that the constellations are all part of a story book in the sky. As the Earth spins round and moves around the Sun the sky changes – it looks as if someone is turning the pages of the story book. The phases of the Moon, the wandering planets and the occasional spectacular comet mean that you never get the same story twice. To the first people on the Earth this made the story book come alive.

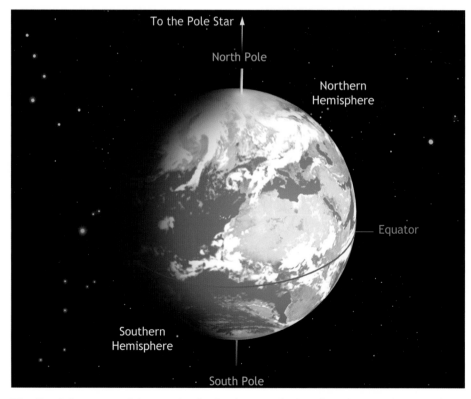

The Earth in space with stars in the background, showing the Northern and Southern Hemispheres and the poles.

The Earth is completely surrounded by a star-filled sky. We can only see the stars at night time. The stars you see at night depend on where you are and what the time is. In the **Northern Hemisphere**, mainly northern stars can be seen. In the **Southern Hemisphere**, mainly southern stars can be seen. Near the **Equator**, both northern and southern stars can be seen. From any place on the Earth, at any time, only half of the stars can be seen. The Earth itself is in the way of the other half!

You don't see the same constellations all night long. As the Earth turns, all the stars look as if they are moving around the sky – but remember, they are not really moving, it's *you* moving around *with the Earth*.

Rising and setting

Some stars **rise** and **set**. On any clear night, try looking to the east and find a star or group of stars that are very low down – as low down as trees or buildings on the horizon. Look again an hour or so later and you will see that the stars have all risen up in the sky. These stars will keep on rising, reach their highest when they are due south (or due north if you live in the Southern hemisphere) and then, eventually, move down to set in the west. Many stars appear to rise and set in this way as the Earth spins around.

As the Earth turns and the night goes by, all the stars move in circles about a point in the sky. This point is called the **pole**. Stars that are far from the pole make bigger circles that can go below the horizon at times – so they rise and set. Stars that are close to the pole make smaller circles. You can see these circles in star trail photographs. If the circles are small enough, it means that the stars never set. These are called **circumpolar** stars because they seem to circle the pole.

At the North or South Poles no stars rise or set, and the pole is directly overhead. At the Equator, all the stars rise and set and the North Pole is due north on the horizon, and the South Pole is due south on the horizon.

Taking photographs of star trails

Star trail photographs like this one are actually very easy to take, if you have the right kind of camera. First let's understand how a camera works. When you take a photograph with a normal camera you hear a click sound. If you listen carefully you can actually hear two clicks. When you press the button to take a photograph this causes a shutter to open and let light into the camera – this is one click. The light falls on the back of the camera and an image is made on the film. If too much light falls on the film the photograph will look bad. It can, for example, make people's faces seem very white. To stop too much light getting into the camera, the shutter automatically closes very soon after it opens – this is the second click.

If you are taking a picture of the night sky the starlight is much, much fainter than a normal picture taken during the day. To get more light into your camera you can leave the shutter open longer. If you leave the shutter open long enough, the star appears to move because the Earth has turned, and will leave a trail on the photograph. One end of the trail is where the star was when the shutter opened and the other end is where the star was when the shutter closed. So to take a star trail photograph, all you have to do is put the camera on a stand, point it at some part of the night sky and open its shutter. Later, you can return and close the shutter, and you will have your star trail photograph on the camera's film. Not all modern cameras let you leave the shutter open, but many older ones do.

If you live in the Northern Hemisphere you can see a star which looks as if it never moves at all. This is The **North Star** or The **Pole Star**. This star is always due north and doesn't get any higher or lower in the sky.

Image opposite. Star trails around the Anglo-Australian Observatory. Stars that make complete circles are called circumpolar stars, all other stars have their circle broken by the ground and so will seem to rise and set.

Constellations

Constellations are patterns of stars in the sky. Over many thousands of years, human beings living on the Earth have looked up and seen the shapes of people, animals and everyday objects – they 'joined the dots' of the stars to form patterns in the sky. They gave these shapes names, some of which we still use today. But why did they bother with constellations at all? One reason is that having constellations makes it easier to find your way around the sky. This is useful for finding north, or working out the date or the time. When there were no compasses, clocks or watches, this was all very important.

There is another reason why constellations were so important. Imagine going back to a time before paper and pencil had been invented. If you wanted to draw a picture, you had to scratch it out on a dark cave wall or write it in sand that could blow away. But at night an amazing pattern of bright specks of light would appear above your head. By joining the dots you could see almost any picture you wanted. You could imagine seeing your favourite shapes in the sky. You might tell others about the constellations you have made up. They might tell others, and your constellation could be passed down from generation to generation for hundreds or even thousands of years. This is how many of the constellations got the names we use.

Constellations and seasons

You don't see exactly the same part of the sky every night, though between one night and the next you won't see much of a difference. Over a few weeks you'll definitely notice that you can see some constellations that you couldn't see before and some constellations that you could see aren't there any more. This is because the Earth is moving around the Sun. As the Earth moves round in its orbit, the night side of the Earth (the side facing away from the Sun) faces out to different parts of space, where there are different constellations.

This means that you see different constellations in different seasons. **Orion** and **Taurus** are (Northern Hemisphere) winter constellations, because you can see them on winter evenings. **Cygnus** and **Scorpius** are (Northern Hemisphere) summer constellations, because you can see them on summer evenings.

WINTER
Orion

Day Night

Night Day

The Sun

SUMMER
Scorpius

As the Earth goes around the Sun during the year, the night time side of the Earth faces out to different parts of space. This is why, for example, Orion is a winter constellation and Scorpius is a summer constellation in the Northern Hemisphere.

Naming the constellations

So who made up the constellations and their names? The earliest people on the Earth were hunters and gatherers. They looked up into the sky and saw shapes that were important to them – like **Orion** the Hunter. Much later in human history, English farmers looked up into the sky and saw the shape of a **Plough**. Russian peasants, looking at the same group of stars, called it **Ursa Major** or **The Great Bear**. People in France called it **Le Casserole** meaning the Saucepan. People in the USA called it **The Big Dipper** meaning a soup ladle. All of these different names are used today. The only people who have decided on one set of names for the constellations are the astronomers. For example, they always call the Plough 'Ursa Major', and never any of the other names. They had to do this so that they all knew what other astronomers across the world were talking about.

If you like, you can look up into the sky, join the dots and make up your own constellations. Tell other people about your constellations. Maybe one day people all around the world will be using one of your constellation names!

Do the constellations mean anything?

Humans have made up all the constellations, so what they mean is only ever inside the head of the person who is looking at them. One person's Plough is another person's bear. Also, the stars in a constellation have nothing to do with each other – they can be very, very far apart, even if they appear to be right next to each other in the sky. Imagine looking up while standing in a street. You might see your hand next to a street light, which is next to the Moon, which is next to a planet, which is next to a star. All of these things are far away from each other, yet they can be next to each other when you look at them.

Star names and magnitudes

The brightest star in a constellation is usually called **alpha**. For example, the brightest star in the constellation of **Centaurus**, the nearest star system, is called Alpha Centauri. Notice that the end of the constellation name is changed too. The second brightest star is called **beta**, and the third **gamma**. The names alpha, beta and gamma come from the old Greek alphabet – in fact that's where the word *alphabet* comes from! It's worth getting to know the symbols of the Greek alphabet:

α is alpha β is beta γ is gamma

There are exceptions to the Greek letter naming rule, usually when many stars in a constellation are about the same brightness. The Plough is one such exception.

The brightest stars in the sky usually have their own names. For example, the brightest star in the constellation of Taurus the bull is **Alpha Tauri**, which is also known as **Aldeberan**.

Astronomers measure the brightness of a star by its **magnitude**. The ancient Arabs introduced this system with six levels of magnitude. The brightest stars were magnitude 1 and the dimmest stars were magnitude 6. Remember that *the bigger a star's magnitude, the dimmer it is* – this can be confusing at first, but you soon get used to it. The Arab astronomers sorted all the stars they could see in the sky (about 6000!) into these six groups by comparing the brightness of each star with the stars next to it.

Modern astronomers have telescopes and devices for measuring the brightnesses of stars very accurately and have updated the magnitude system. The dimmest stars we can see without a telescope under perfect conditions (no streetlights, no haze) are still magnitude 6. But, the brightest stars in the sky can have a magnitude of less than 1. **Vega** has a magnitude of 0 (zero) and the brightest star in the sky – **Sirius** – has a magnitude of –1.5 (minus one and a half). Telescopes or binoculars can let you see stars that are dimmer than magnitude 6. The best telescopes we have now can probably see stars as dim as magnitude 30. Even a small telescope could probably show you stars as dim as magnitude 10.

The Hubble Space Telescope (HST) orbits the Earth and lets us see some of the dimmest stars in the sky. Many of the pictures in this book came from the HST.

The zodiac

You've probably heard of many of the constellations in the zodiac. They are Pisces, Aries, Taurus, Gemini, Cancer, Leo, Virgo, Libra, Scorpius, Sagittarius, Capricornus, and Aquarius. These constellations are special because the planets and the Moon appear to move through them. The Sun also moves through each constellation of the zodiac, but you can never see the constellation the Sun is in because it will be daylight when that constellation is up in the sky – think about it.

The zodiac has a special meaning in astrology. People often confuse Astronomy with **Astrology**. Astronomy is a science. Astronomers, who are scientists, look at something in the sky, such as a star, and come up with ideas about it, like how hot it is. They call these ideas **theories**. After coming up with a theory, they will try and check it, by, for example, looking to see if a theory for one star is true for others. If they were wrong, they come up with new ideas. Astrology is not a science. Astrologers believe that patterns of the Sun, Moon and planets in the constellations can predict our lives. They believe this, and have some theories, but most astrologers aren't interested in testing their theories.

Touring the stars

You don't need a starship to explore the stars. You don't even need a **telescope**. The dark side of the Earth – where you are every night – gives an excellent view of the Universe and its stars. You can start with your eyes, a dim flashlight, some warm clothing and a **star map** or **planisphere**. A planisphere is a handy little round star map with a window that moves round to show you what you can see at a particular date and time. You can buy star maps and planispheres from most bookshops, though you might have to order them.

In this book we've given you a description of what we think are the most interesting constellations. We've numbered them so that if you've found number 1, number 2 won't be too far away from it. If you are observing in the Northern Hemisphere in winter – say in Europe, North America or Russia – try starting at number 1. If you are in the Southern Hemisphere try starting from the last number and work backwards. If you want to know exactly what you can or cannot see, use a planisphere, or even better a computer **planetarium** (there are many good free ones on the internet).

Once you know the constellations in this book, you will be able to find other constellations for yourself. At first you may find it difficult to find your way around the night sky, but as you get to know more constellations, you will find it easier.

Make sure you prepare for your observing session by following the checklist. If you can, get everything ready before you go out. That way, you won't waste time when you are out under the stars.

Checklist for observing

1. **Look at your star map beforehand.** Get to know the constellations you want to try and find. Remember that the real night sky has no lines, no handy names and no compass bearings on it.

2. **Wear warm clothing.** It can get very cold at night. Your hands and feet will get cold first, so two pairs of socks and gloves are important. You may need a few layers of clothes and a warm hat.

3. **Use a dim, red flashlight.** You will need to look at a star map quite often. If you use a bright flashlight you will dazzle yourself. Use the dimmest flashlight you can find. If you can, put some clear red plastic on it because red light does not dazzle you as much. It takes several minutes for your eyes to adjust to the darkness, then you can begin observing properly.

4. **Find a good spot.** Just outside your house may be a good place to start. You want to be away from as many streetlights as possible, but have a good view all around you. If you can, ask your parents and neighbours to draw their curtains or close their blinds and turn off their outside lights. You want to make it as dark as possible.

5. **Keep safe.** Make sure your parents know where you are and what you are doing. Better still, take an adult out with you, and point out all the stars and constellations that you have found.

1 The Plough

This constellation gets its name from the old horse-drawn plough, which was a much more common sight a hundred years ago than it is today. A more recent name that came from the USA is The Big Dipper, a dipper being a kind of soup ladle.

The whole constellation – not just these seven stars – is called Ursa Major or The Great Bear. Since all these stars are close to The Pole Star, the Plough will never set for many observers in the Northern Hemisphere (northern Europe, northern America and northern Russia). To find The Pole Star, imagine that these seven bright stars are in the shape of a saucepan with a long handle. Find the two stars at the front of the pan part of the saucepan and imagine a line joining these two stars (which are sometimes called **The Pointers**). Follow the line up and out of the saucepan and you will find The Pole Star. (Look at the north polar area map on page 20.) If you do this again an hour or two later, you will see how The Pole Star has stayed still but the Plough has moved round.

If you look carefully at the second star from the end of the Plough's handle, you will see that it is actually a double star. The brighter of the two stars is called **Mizar**, and the dimmer one is called **Alcor**. They are the easiest to see of all double stars in the sky. If you can't see them then you might have to get your eyes tested!

The brightest seven stars in the constellation of Ursa Major (The Great Bear) are known as the Plough. The two stars in its 'pan' point to the Pole Star in the constellation of Ursa Minor (The Little Bear).

The Pointers

To the Pole Star

Mizar

Alcor

Ursa Major

The constellation of Ursa Major.

2 Cassiopeia

If you have found the Plough, you should be able to find another important constellation of the Northern Hemisphere – **Cassiopeia**. It is on the other side of The Pole Star from the Plough. In Greek stories Cassiopeia was a proud queen, the mother of **Andromeda**. In the sky Cassiopeia is in the shape of a big W or M.

To find Cassiopeia, start at the end star of the Plough (**Alkaid**). Imagine a line joining Alkaid to The Pole Star. Continue the line through The Pole Star for about the same distance again and you will come to Cassiopeia.

The **Milky Way** runs through Cassiopeia. It must be very clear and dark to see the Milky Way – if you are in a city, you almost certainly won't see it. The Milky Way is made up of the billions of stars in our galaxy. Astronomers think our galaxy contains over one hundred billion stars.

Cassiopeia

Cassiopeia is a constellation close to The Pole Star, but on the other side from Ursa Major (the Plough). There are lots of stars here, but if you take a moment to look carefully you will see all the stars in the diagram.

The constellation of Cassiopeia.

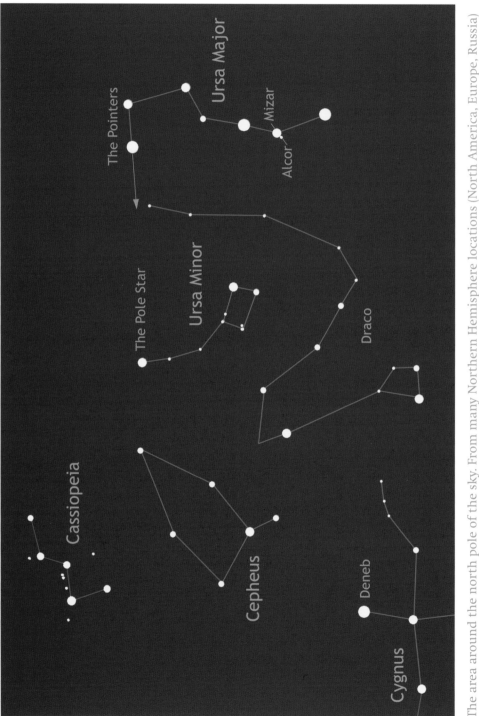

The area around the north pole of the sky. From many Northern Hemisphere locations (North America, Europe, Russia) all these stars are circumpolar, that is, they never set, but just move in circles around the Pole Star. Shown here are a few fainter constellations: Ursa Minor (The Little Bear, containing the Pole Star), Cepheus (a King married to Cassiopeia in Greek myths) and the long, twisting constellation of Draco (the dragon).

The Pointers

Ursa Major

Mizar

Alcor

The Pole Star

Ursa Minor

Draco

Cassiopeia

Cepheus

Deneb

Cygnus

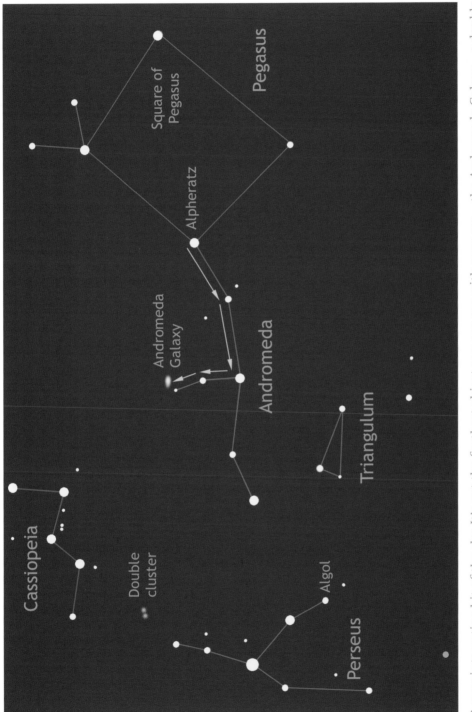

A very interesting bit of the sky. It's got the furthest object you can see with your eye – the Andromeda Galaxy – a double star cluster, and the variable star Algol. In addition to all these really interesting things, there is also the most boring constellation in the whole of the night sky: Triangulum – the Triangle!

3 Perseus

Perseus is a T-shaped constellation right next to Cassiopeia. If you look carefully you might be able to see a fuzzy patch between the two constellations. If you look at it with binoculars you can see that this is in fact two **star clusters**. (You'll see an even better example of a star cluster in the constellation of Taurus.)

The star at the bottom of Perseus's 'T' is called **Algol**. Algol is a **variable star**, which means that it changes in brightness. You won't be able to see it getting brighter and dimmer before your eyes, but you can tell that its brightness is changing from night to night. In roughly three days it will go from its brightest, to dimmest and back to its brightest again. If you watch it for three nights in a row at about the same time, and compare its brightness with nearby stars, you should be able to tell that Algol has changed in brightness.

There are many interesting things to see in and around the constellation of Perseus: for example, the variable star Algol and the double star cluster between Perseus and Cassiopeia.

The constellation of Perseus.

23

4 Andromeda

If you imagine drawing a line from The Pole Star through the middle of Cassiopeia, you'll arrive at the constellation of Andromeda. Andromeda isn't a particularly interesting constellation to look at because it doesn't have any bright stars. But – and this is a big but – there is something very special in Andromeda. The furthest object that the human eye can see is in Andromeda – The Andromeda Galaxy. To see it, you'll need a really good, clear sky and you will need to be away from any cities or towns with lots of bright streetlights.

To find the Andromeda Galaxy, you first have to find **The Square of Pegasus** (Pegasus is a constellation itself). The square has four stars, one at each corner. (See the map on page 21.) The stars are about as bright as the stars in The Plough and are about as far apart as your knuckles held at arm's length. Now find the star in the square called **Alpheratz** (shown on the map) and follow the stars joined by the white lines on the map. In the real night sky there are no white lines, but there's an easy way to remember the directions. From Alpheratz, move two stars along towards Perseus, then 'turn right' and move along two more stars. The Andromeda Galaxy will look like a faint smudge, but remember, you are looking at something over two million light-years away, and looking over two million years back in time.

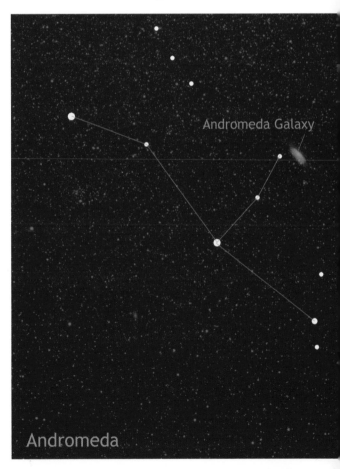

This photograph shows the constellation of Andromeda and The Andromeda Galaxy.

Andromeda and Andromeda Galaxy.

5 Cygnus

Cygnus is the constellation of the swan. It is shaped like a cross, and its brightest star, **Deneb**, is in the tail of the swan. If you have dark enough skies you can see that the swan is flying along the Milky Way, as shown in the photograph. Deneb is one of the three very bright stars that make up the **Summer Triangle** (which can be seen in Northern Hemisphere summers). The other two stars are **Vega**, which is in the constellation of **Lyra** (the lyre, an ancient musical instrument) and **Altair**, which is in the constellation of **Aquila** (the eagle). The bright star at the bottom of the photograph is Vega.

Deneb, the brightest star in the constellation of Cygnus, together with Vega and Altair (not shown here), make up the Summer Triangle which dominates the summer skies.

Cygnus is 'flying' along the Milky Way with Deneb, its tail, being to the top and right in the photograph. Vega, the brightest star in the constellation of Lyra is at the bottom of the photograph.

6 Hercules

Hercules is the name of a warrior, and the constellation shows him with his arms outstretched running into battle. His body is a squint square of four stars but he doesn't seem to have much of a head. On a clear, dark night you can see a little fuzzy patch below the right armpit. This is a special kind of star cluster called **M13**. It is a group of millions of stars all held together in a ball. Even a small telescope can show you this spectacular sight.

The globular cluster M13 – a ball of stars in space held together by gravity.

The constellation of Hercules. In the constellation of Hercules you can find the globular star cluster called M13 – a good first object to find with binoculars or a small telescope.

Hercules and M13.

M13

Hercules

7 Taurus

If you follow an imaginary line from the middle of Cassiopeia through the middle of Perseus you will come to the constellation of Taurus (the bull). Taurus does not have an obvious shape. The brightest star in Taurus is **Aldebaran**. It is an orange-red colour, though it is not quite as bright as Betelgeuse in Orion. Aldebaran is also called 'the eye of the bull'.

If you look around Aldebaran you'll notice that there are lots of faint stars. This is a star cluster called the **Hyades**. Unlike a constellation, these stars are all close together and were born about the same time.

You can find another star cluster – the famous **Pleiades** or **Seven Sisters**. There are more than seven stars there, but you'll be lucky to see even seven without binoculars or a telescope. These stars are all about 60 million years old. They were born about the same time that the dinosaurs died out on the Earth.

The Pleiades star cluster.

The constellation of Taurus contains two star clusters: Hyades loosely spread around its brightest star Aldebaran, and Pleiades, which is sometimes called the Seven Sisters.

The constellation of Taurus and its famous star cluster – Pleiades.

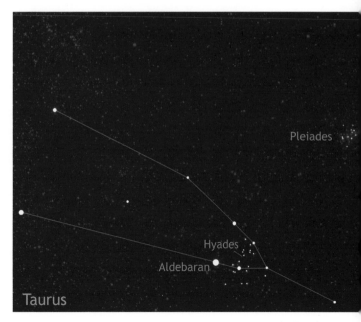

Pleiades

Hyades

Aldebaran

Taurus

8 Orion

Orion is one of the easiest constellations to spot and is easier to find than The Plough or Cassiopeia, but it is not always above the horizon – it rises and sets. Orion is a spectacular sight in the evening during winter in the Northern Hemisphere.

Orion has the shape of a man. He has two stars for his feet, three stars in his belt, two stars for his shoulders and some dim stars make up his head. Orion is a hunter, who is hunting Taurus and Taurus the bull is charging at Orion.

On one shoulder is the bright orange-red star called Betelgeuse. This is an old Arab word that means 'soldier's armpit'. Betelgeuse (some people call it 'beetle juice') is one of the largest types of stars called **supergiants**. It is truly enormous – you could fit a million stars like our Sun inside Betelgeuse. If it was placed where our Sun is, it could swallow up the entire path of the Earth around the Sun!

The Orion Nebula.

Orion, with its bright stars Rigel and Betelgeuse and The Orion Nebula below its belt, dominates the winter sky. Near Orion you can see the brightest star visible in the night sky – Sirius.

The constellation of Orion.

Gemini

Taurus

The Pleiades

Canis Minor

Procyon

Betelgeuse

Orion

Rigel

Sirius

Canis Major

Lepus

Most people think that this is the most beautiful region of the night sky. It's got the brightest star in the night sky – Sirius – the star cluster Pleiades – and the bright stars and nebula of Orion.

On Orion's foot you can see the bright blue-white star called **Rigel** (pronounced rye-jel). It is about 900 light-years away from the Earth. It is a huge star compared to the Sun but is much smaller than Betelgeuse. It is a very powerful star which is 60 000 times brighter than the Sun, but it is very far away.

If you look below Orion's belt, you should see three dim stars in a line, pointing down from the belt. This is **Orion's sword**. The middle of these stars will look slightly fuzzy – you are looking at **The Orion Nebula**, a cloud of gas and dust, lit up by stars inside the nebula. Some of these stars have only recently been born (a few million years ago). There are many nebulae in the sky, but The Orion Nebula is the most famous. If you look to Orion's left, you can see a faint arc of stars that make up his bow.

When you have found Orion, the Hunter, you can find his two dogs following him. Look at the map above. Follow his belt down to your left. The very bright star you can see is called **Sirius** in the constellation called **Canis Major** (the big dog). It is the brightest star in the sky – after the Sun! Canis Major was Orion's senior dog, so Sirius is often called the **Dog Star**.

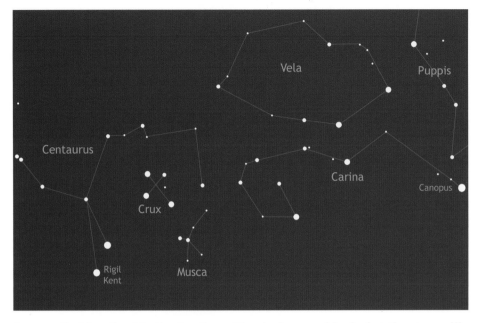

Anyone that has lived in the Southern Hemisphere and looked at the stars will know the Southern Cross or Crux. In the same part of the sky are the second and third-brightest stars in the sky: Canopus in Carina and Rigil Kent in Centaurus. This map shows some other constellations, mostly made up of dimmer stars! Musca (the fly), Vela (the ship's sail), some of Puppis (the ship's stern). If you piece together the constellations of Carina, Vela and Puppis you might see the ship in the sky that sailors saw as they explored the unknown seas of the Southern Hemisphere hundreds of years ago.

Sirius is not alone. It has a small companion (a **white dwarf**) which cannot be seen without a telescope, because it is lost in the glare of Sirius. This small companion of the Dog Star is often called the **Pup**.

Canis Minor (the little dog) is above and left of Sirius. Its brightest star is **Procyon**, another of the very bright stars. It also has a small companion star, but you can't see it without a telescope.

The following constellations are best seen from the Southern Hemisphere, and will be difficult or impossible to see from many places in the Northern Hemisphere.

9 Scorpius

This constellation is the scorpion. It is in the zodiac and is called Scorpio in astrology. For places in the Northern Hemisphere of the Earth, like Britain or northern parts of the USA, Canada and Russia, Scorpius is a summer constellation and doesn't get very high in the sky. Its most striking feature is the bright, red star called **Antares**. According to legend the scorpion stung Orion's foot, which turned bright and blue – the star Rigel.

Antares

Scorpius

The constellation of Scorpius. The greenish glow at the bottom right of the picture is caused by streetlights.

Scorpius, the constellation of the Scorpion, has bright, red Antares as its brightest star.

The Large Magellanic Cloud.

10 Carina and the Magellanic Clouds

Carina is a big constellation, but can easily be found because its brightest star – **Canopus** – is the second-brightest star in the night sky (see page 35). Not too far from Canopus (within about a knuckle at arm's length) you can see **The Large Magellanic Cloud**. This fuzzy patch of sky looks a bit like the Milky Way, and so needs good clear, dark skies if you are to see it. Nearby, you'll also see the **Small Magellanic Cloud** (both clouds are off the edge of figure on page 35). Of course, these aren't clouds at all, but are groups of very many stars that are extremely far away. In fact The Magellanic Clouds are little galaxies just outside our own galaxy, the Milky Way.

The Small Magellanic Cloud.

11 The Southern Cross (Crux)

People who live in the Southern Hemisphere of the Earth cannot see The Pole Star because it is in the northern part of the sky. Since there is no South Pole star, they have no easy way of finding out which way is north or south from the night sky. If you were lost at night in Australia, the constellation of The Southern Cross would be your best bet for finding your way. It has four prominent stars that form the cross, and the long part of the cross always points towards the South Pole of the sky. If you know this, and you've got a bit of experience in judging distance across the night sky, you can find south. The proper name for the Southern Cross is **Crux**.

Crux (The Southern Cross) and Carina with the Milky Way running inbetween them.

The constellations of Crux (The Southern Cross) and Carina.

12 Centaurus

On the other side of the cross from Carina you will find the constellation of Centaurus. According to legends, a centaur is a creature that has the body of a horse, and the chest, arms and head of a man (where the horse's head would be). You've really got to use your imagination to see all that!

The brightest star in Centaurus is Rigil Kent or, as it's also known, Alpha Centauri. Not only is it famous for being the third-brightest star in our night sky, it is also the closest star after the Sun. This is of course no coincidence, as stars will seem brighter if they are closer to us – the Sun is the best example of this.

Alpha Centauri is actually a triple star system. It has two main stars that are both very bright and close to each other, and a third, much dimmer, star. This third star is called **Proxima Centauri** because it is actually the closest of the three stars to us.

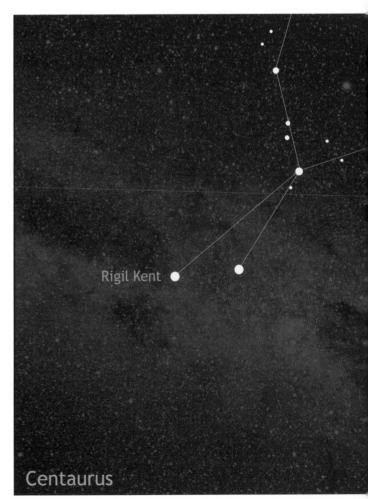

Rigil Kent

Centaurus

The constellation of Centaurus and the Milky Way.

Centaurus. Rigil Kent, also known as Alpha Centauri, is the third brightest star in our night skies.

You thought you saw a star move?

Most people think they have seen a star moving when they see a **shooting star**. This looks like a star moving very fast, perhaps taking less than a second to move across a large part of the sky. A bright one might even leave a trail behind, or a flash of different colours. This is not a star – the name 'shooting star' is very misleading. It is actually a piece of rock that came from space burning up in the Earth's atmosphere. If the rock is small, like a grain of dust, it burns up completely. This is called a meteor. If only some of the rock burns up and the rest lands on the Earth somewhere, it is called a **meteorite**. If this was really a star falling to the Earth (which could never happen) not only would you be blinded by its brightness, you would be burnt to cinders by its heat.

Perhaps you have seen a star move slowly, taking many minutes to travel across most of the sky. It may even have appeared or disappeared. This is almost certainly a **satellite** that we (human beings) have put into orbit around the Earth. We can see them because they reflect light from the Sun. They can disappear as they turn round or if they move into the Earth's shadow. Satellites can be very bright and are becoming more and more common. It is worth checking the internet, newspapers or magazines to see if you have seen a particular satellite – you might even have seen the Space Shuttle!

You may have noticed a bright star that seems to have moved slightly over a few days or weeks. This is almost certainly a planet and probably one of the inner planets, which are always quite near to us. The best way to find out which planet you have seen is to look at a star map in a magazine. Better still, you can also use a planetarium program on a computer to do this, or best of all, visit an actual planetarium. You can also recognise planets by their colour – the easiest one is Mars, which is orange-red. One handy hint for spotting planets is that they don't twinkle as much as stars.

If you find a bright star that isn't on any star maps and doesn't seem to be one of the planets, you may have found a nova, or even a supernova, not too far away from the Solar System. These are not new stars, or bright stars that have moved – they are dim stars that have suddenly got much brighter after a big explosion. If such a bright nova or supernova goes off, you would be sure to see it in the headlines of the news.

Perhaps you have looked up at the stars through tree branches or railings and noticed them move by. *The stars are not moving – you are.* In fact, we all are. As the Earth turns we all move around and the stars, and everything else, will rise and set.

You may have noticed that in winter you see different constellations in different places from spring, summer or autumn. This is because the Earth is moving around the Sun, so that at different times of the year we look out at different sets of stars at night time.

The main thing to remember is that you
will never see a star move.

As a rule, only things that are quite near us can easily be seen to move over the time of a human life and all of these things are in the Solar System.

The Solar System

What does it mean?

The word *sol* is the Latin word for 'Sun'. The **Solar System** means the **Sun** and everything which is in orbit around it. This includes **planets**, **moons**, **asteroids**, **meteoroids** and **comets**. The Sun's **gravity** pulls on all these things and stops them flying off into space.

How do we know about the Solar System?

Early people must have wondered about the sky and the Earth. We think they were very much like us in the amount of intelligence they had. They had no telescopes, televisions, satellites, planes or cars – and no science. They believed that the Earth was flat and that, if you journeyed far enough, you would fall off the edge. We still talk about 'the four corners of the Earth' and about travelling 'to the ends of the Earth'.

Of course, we all know now that the Earth is a huge **sphere** (ball-shape) because we have seen pictures of what the Earth looks like from space.

Next time you are in the countryside, try to imagine that you are one of the early people trying to make sense of your world. You would surely believe that the Earth was flat – except for some bumps that appear here and there (hills and mountains). The sea would look even flatter. You would probably be frightened to go far from land just in case the sea suddenly stopped and you fell off the edge into some underworld!

If anyone had told you that the Earth was a sphere, you would have laughed at them and thought they were crazy.

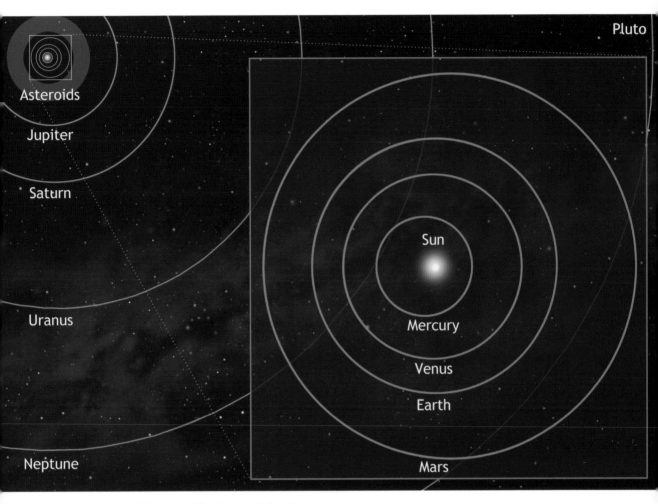

Layout of planet orbits in the Solar System.

How did people work out that the Earth is a sphere?

Here is one way it could have happened. Imagine yourself standing on land looking out to sea on a clear day. The horizon (where the sea seems to meet the sky) is a clear line. Suddenly, you notice a funnel sticking out of the sea. The funnel then appears to get bigger, and you see more of it until the ship's hull starts to appear. It's as if the ship is rising out of the

An astronaut orbiting the Earth is in no doubt that it is a sphere.

sea as it comes towards you. Of course, the ship isn't rising out of the sea at all, it's just that the curved surface of the Earth hides the bottom of it from you until it gets to about ten miles of where you are standing. A ship sailing away from you appears to sink into the sea for the same reason.

From where we are on the Earth, the Earth itself appears to be the centre of the Universe, with all the stars, planets, Sun and Moon revolving round it. It took many thousands of years to work out that the Earth and planets actually move around the Sun and that the Earth itself turns round once every day. Around 290 BC, a Greek called Aristarchus suggested that the Sun was at the centre and that all the planets, including the Earth, go round the Sun, but everyone else thought it was nonsense. It took nearly 1800 years before people realised that Aristarchus had got it right.

A ship appears over the horizon because of the Earth's curvature. You see only the funnel of a ship that is many miles away. The top half of the ship now appears above the sea surface, which is curving away from you. When the ship is only a few miles away you can see all of it.

The night sky

Imagine once again that you are one of the early people on the Earth. Stand outside on a clear cloudless night and look up. What would you see? A dome of darkness, with little points of light that form patterns. If you came back a few hours later, you would see that the patterns were the same but they had moved round to the west as though the dome had been pushed round.

People began to realise that the stars were just like our Sun – the star at the centre of the Solar System. Where did our Solar System come from? Has it always been there or did it have a beginning?

The next chapter will tell you about the Universe and how scientists think it began. Most of them think it started about 12 billion years ago. Our Solar System is a tiny part of the Universe, although it seems enormous to us. The Solar System did not suddenly appear when the Universe began. It is less than five billion years old.

Gravity

Think of any two objects. They can be anything you like, for example: a knife and a fork, a table and a chair, a tree and an apple, the Earth and the Moon, or the Sun and another star. Each and every pair of objects is being pulled together by what we call **gravity**. In fact, everything in the Universe is pulled towards everything else in the Universe by gravity. So it is true to say that the Earth's gravity pulls on the table, or the Moon's gravity pulls on the fork, and even that the apple's gravity is pulling on the Sun!

You know that forks don't fly off to the Moon, and that neither apples, nor anything else on Earth, cause the Sun to crash down on us. The reason these things don't happen is that the strength of gravity's pull depends on two things.

The first is the **mass** of the object. The apple is very small, and doesn't have much mass, so its pull on the Sun is absolutely tiny, certainly much smaller than the pull of all the planets. The Earth is more massive (has more mass) than tables, trees or apples, so almost everything in the everyday world around is pulled towards the Earth. That's why apples *fall* from trees.

Now, you might know that the Sun is much bigger than Earth and is much more massive. So why don't apples fly off towards the Sun? The reason is that the pull of gravity also depends on the **distance** to the object doing the pulling. Although the Sun has much more mass than the Earth, we are much closer to the Earth, so we feel its gravity more.

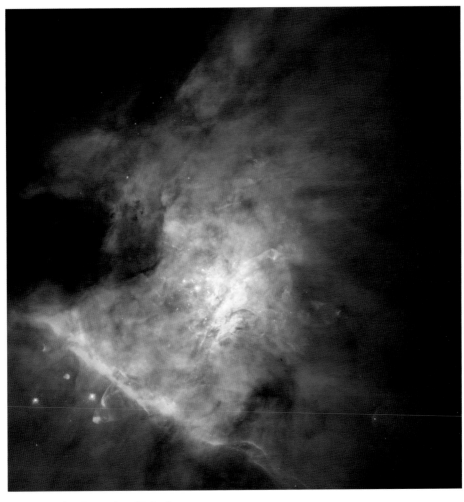

Starbirth in the Orion Nebula.

A star is born

We are about to go on a journey back in time about 4½ billion years to the time when the Solar System was formed. Of course, there were no people alive at this time because there was no Earth. How can we work out what happened when no one was there to see it?

Space between the stars is not quite empty. It contains **hydrogen atoms** (tiny particles that make up the gas hydrogen) and particles of dust. Astronomers, using telescopes, can find areas in space where the gas and dust are much thicker than usual – a **nebula**. They can see that stars are often born in these regions of space. In fact these baby stars inside the nebula light up all the dust and make the gas glow.

Here is a description of how some astronomers think our Solar System was born. A star near a nebula exploded. This caused the cloud of gas and dust in the nebula to become all squashed together. Then, this cloud's own gravity pulled the outside of the cloud towards the middle, making it shrink. As it shrank, it began to **rotate** (spin round) faster and faster. Most of the gas and dust ended up in the core, or centre, of the spinning cloud. As the gas and dust became more and more tightly packed together, the core became hotter and hotter. The inside became hot enough to cause **nuclear reactions** in its centre. The core began to shine, giving out heat and light energy – a star was born. It sent waves of energy outwards which blew some of the gas and dust out into space.

This gas and dust began to lump together. The lumps kept spinning around the new star and started to gather together, making bigger lumps. After a long time, the lumps became very big. They had become planets. Smaller lumps became moons. Even smaller lumps became asteroids, comets or meteoroids – which we will come to later. Remember, we do not yet know for sure that this happened. Scientists must always be ready to admit they are wrong when they discover something new. If you became an astronomer, you could one day be the one to find out that all the older scientists were wrong about how the Solar System came about!

The birth of the Solar System. A star explodes. (Top left.) Nearby a gas cloud is squashed, and gravity pulls this inwards; as it does this the gas starts to heat up. As the gas cloud shrinks, it spins round faster and faster and spreads out into a disk. At the centre the gas is hottest and a star – the Sun – starts to shine. Gravity causes clumps of gas and dust to appear in the spinning disk. These clumps cool and planets start to take shape. (Bottom.)

The Sun, our star

There would be no life on the Earth at all without the Sun. It provides the **energy** which everything needs in order to be alive. If you were able to journey far out into space and look back at our Sun, you would see that it is just one of many stars. Stars come in different sizes from **supergiants** to **dwarfs**. Our Sun is an ordinary medium-sized star. It is not very special compared to other stars but, for us on the Earth, it is the most precious star in the Universe because without it, we would not exist.

It is not hard to understand why many ancient peoples thought the Sun was a god. The Sun gave light and warmth and made the crops grow. Without the Sun, the Earth would be a cold lifeless world – no one would survive.

All the energy that is produced by the Sun constantly pours out of its surface as light. In fact, the Sun produces so much **light energy** that even though the Sun is 150 million kilometres away from the Earth, your eyes can be badly damaged by looking directly at the Sun.

> ⚠ **Never look directly at the SUN,**
> **and never, never with binoculars or a telescope.**
> **You can be blinded for life by looking at the sun.**

If the Sun is a star, why is it so bright?

This is because it is so near to the Earth. It is about 150 million kilometres away. This seems a very big distance, but it is tiny if you compare it to the distance to any of the other stars. All the other stars are so far away that we see them only as tiny points of light when the sky is dark. During the day, the light from the Sun drowns out the light from the other stars so that we only see the other stars at night time.

What is the Sun like? How does it work?

The Sun that you see in the sky looks like a circle. It is sometimes drawn as if it has lots of rays coming out of it. These rays have nothing to do with the Sun and are caused by the air and water in the Earth's **atmosphere**. The bright circle that you see in the sky is the surface of the Sun. The Sun is a sphere which means it is ball-shaped and, like all balls, has a circle shape no matter which way you look at it.

The Sun is enormous. Everything else in the Solar System – the nine planets, asteroids, meteoroids and comets – could fit inside the Sun if it was hollow. Like the other stars, the Sun is a sphere of extremely hot gas. It is made up of mainly two types of gas: hydrogen and helium.

The Sun is made up of layers. It has a **core** (a sphere in the middle) that is bigger than **Jupiter**. This is where the Sun's massive energy is produced. The core is at a temperature of about 15 000 000 °C. The huge temperature and pressure in the core forces bits of hydrogen together, turning them into helium. This is called **nuclear fusion**. This nuclear fusion goes on all the time and makes enormous amounts of energy.

This energy takes a long time to reach us. Light can spend millions of years bouncing around inside the Sun until, one day, the light reaches the surface and travels through the Solar System. Eight minutes after leaving the Sun that little bit of light will arrive at the Earth. In fact, it could come in through your bedroom window and be what wakes you up one morning.

Because the Sun has no solid surface and is made up of very hot gas, there are many things happening on the Sun that could never happen on the Earth. You can get **sunspots**, which are dark, cool patches on the Sun. Although you can't see it easily, the surface of the Sun is constantly bubbling because of its heat, like a pan of boiling water. The surface of the Sun – the bright circle we normally see – is called the **photosphere**. The temperature of the photosphere is 6000 °C. This is much hotter than the hottest oven you'll ever find on the Earth. A hot day on the Earth would be 20 to 30 °C – the Sun is not a pleasant place for humans to go to!

Outside the photosphere is the **corona** – the outer atmosphere. It stretches for millions of kilometres into space. It is made up of gases which are extremely hot. The temperature of the corona is over 2 000 000 °C. No one is quite sure why it is that hot, so many solar astronomers are working hard at the moment to try and figure it out.

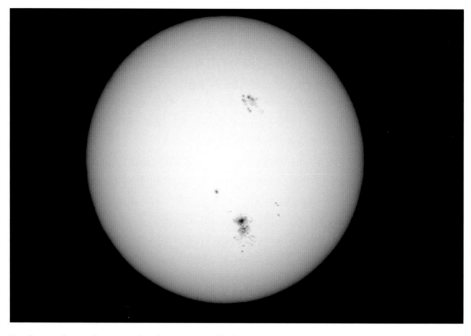

Dark, cool patches on the Sun are called sunspots.

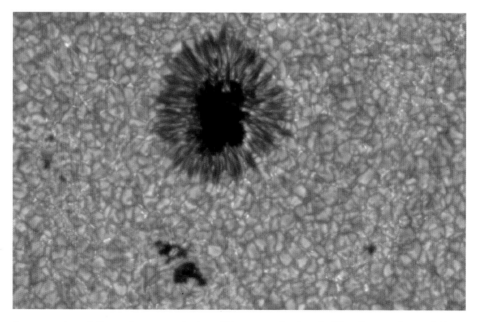

A sunspot and the solar granulation. The granulation is like the surface of the Sun bubbling as hot gas reaches the surface of the Sun, sinks and then cools.

Normally you cannot see the corona because it is lost in the glare of the photosphere. You can only see it during a **total solar eclipse** – when the Moon goes in front of the Sun and covers the photosphere.

The Sun sends a wind out into space, called the **solar wind**. It takes a few days to reach the Earth. When there are 'storms' in the atmosphere of the Sun, the solar wind can have strong 'gusts'. Although we don't feel them, they can cause power cuts, damage the electronics in satellites and even harm astronauts.

A total eclipse of the Sun where the Moon blocks out the Sun's bright photosphere and lets us see its faintly glowing corona.

What is happening in the atmosphere of the Sun can also have a big effect on the Earth and its atmosphere. It doesn't give us good or bad weather, or make it rain on a particular day, but some scientists think it causes patterns in our weather. Hundreds of years ago, when the Sun had few sunspots, scientists think parts of the Earth went into a little ice-age.

Sometimes a violent storm on the Sun can send gas racing out into space, and sometimes this will hit the Earth. Most of the time, it doesn't cause too much damage, but it might cause an **aurora** – which we can see. This is a beautiful display of slowly-changing colours that glow in the sky. If you live in the Northern Hemisphere you can see the **aurora borealis** or **northern lights**. If you live in the Southern Hemisphere you can see the **aurora australis** or **southern lights**.

An aurora photographed from the ground.

An aurora as seen by astronauts onboard NASA's space shuttle *Discovery*.

Why doesn't the Sun blow itself up?

Fortunately for us here on the Earth, the Sun is kept balanced by the forces of gravity which pull inwards and the pressure of gas which pushes outwards. (You can feel the pressure of a gas by squeezing a balloon.) So the Sun will not blow itself up like a bomb because its gravity holds it together. But, once the Sun runs out of energy, the gas won't be able to balance gravity and the Sun will eventually shrink – but that won't be for a long, long time – five billion years in fact.

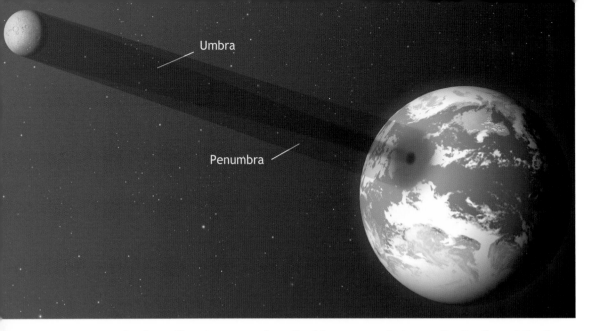

Umbra

Penumbra

A solar eclipse happens when the Moon moves between the Earth and the Sun.

Solar eclipses

A solar eclipse happens when the Moon comes directly between the Earth and the Sun, blocking out light from the Sun. By an odd coincidence, the Moon is exactly the right size to just cover the Sun's disc because the Moon, although much smaller than the Sun, is closer to the Earth.

You have to be at a particular place on the Earth, on a particular date and time, to see a total eclipse of the Sun. A total solar eclipse allows us to see the Sun's upper atmosphere (the corona) and can be a spectacular sight. Many people travel thousands of miles to be in the right place at the right time for a total solar eclipse.

Will the Sun die?

In the distant past, people thought that the Sun had always been there and would last forever. Nowadays we know this is not true. **Geologists** (scientists who study rocks) have discovered that the oldest rocks in the Solar System are about 4½ billion years old and they have shown that these rocks are as old as the Sun itself. This means that the Sun was born about four and a half billion years ago. Some stars live for one million years. Some will live for 100 billion years! Our Sun is an average star with a lifespan of about ten billion years, so right now it is middle-aged. Since it has about five billion years of life left, we shouldn't worry too much about it. Have a look at the calendar of the Universe to see how old the Sun and the Solar System really are.

January	February	March	April
Big Bang **First stars** **First galaxies**		**Milky Way forms**	

May	June	July	August
			Solar System forms

September	October	November	December
First lifeforms			

December 10	First animals	
December 25	First mammals	
December 29	Dinosaur extinction	
December 31 11.30:00 p.m.	First humans	
December 31 11.59:00 p.m.	First civilisations	
December 31 11.59:59 p.m.	Humans on Moon	

A calendar showing the history of the Universe. The scale is such that 1 month is equivalent to 1 billion years, 1 day is 33 million years, 1 hour is 1.4 million years, 1 minute is 23 000 years, 1 second is 385 years.

The nine planets

The planets are the next biggest things in the Solar System after the Sun. They are all spheres (ball-shaped) and go around the Sun in paths that are a bit like circles – the path of a planet around the Sun is called an **orbit**. We say that the planets 'circle' the Sun but the orbit of each planet is like a flattened circle – an **ellipse** or oval – and is called an elliptical orbit. Since the orbits are almost circles, each planet stays at roughly the same distance from the Sun as it moves around. The Sun's gravity (pulling force) is strong enough to keep the nine planets of the Solar System in orbit with the Sun at the centre.

The planets nearer to the Sun, like **Mercury** and **Venus**, feel the pull of the Sun's gravity much more than the planets further away, like **Pluto**. The nearer a planet is to the Sun, the faster it travels in its orbit, and the shorter the distance it has to move in one orbit. Mercury, the nearest planet to the Sun, takes 88 days to orbit the Sun once. The Earth, which is further away from the Sun, takes 365 days, which we call one year. Pluto, the furthest away from the Sun, takes over 248 Earth years to make one orbit.

As the planets orbit the Sun, they also spin – like a football spins as it flies through the air. The Earth takes about 24 hours to spin once, and that is why a day is 24 hours long. Remember, the year is to do with a planet's orbit, the day is to do with a planet's spin.

The four planets nearest to the Sun are Mercury, Venus, Earth and Mars. They are called **terrestrial** planets. They are made of rock and have a surface you can stand on. The next four planets are huge compared to the terrestrial planets. They are **Jupiter**, **Saturn**, **Uranus** and **Neptune** and are called **Jovian** planets. These planets are mostly made of liquids and gases. If you tried to stand on any of the four Jovian planets, you would just sink down inside and eventually become part of the planet. The larger planets have atmospheres that are layers of gas kept near the surface by the planet's gravity. Smaller planets and most moons have lost the atmosphere they once had because their gravity was too weak to hold on to the gas.

Pluto, the ninth planet, is the odd one out. It doesn't seem to fit into either of these two groups of planets. Very little is known about it and no spacecraft has ever landed there, though there are plans to send one there within the next 20 years.

The planets are all different. We know a lot more about them today because of the space probes that have either visited the planets or passed close by them. These space probes carry no astronauts, being robots that send information back to the Earth. They have to travel for many years to reach the other planets. Voyager 1 and Voyager 2 were launched in 1977. They have sent back information about Jupiter, Saturn, Uranus and Neptune.

Voyager 1 has travelled further from the Earth than any other made object. It is now more than 10 billion miles away, further than Pluto, and its radio signals take almost 10 hours to travel back to the Earth.

All nine planets orbit the Sun in the same direction. Imagine looking down on the Solar System from above the Earth's North Pole. All the planets would orbit the Sun in an anti-clockwise direction.

The Voyager spacecraft.

Most planets have moons – sometimes also called satellites (a satellite is anything that goes around a planet, whether it was made by humans or not). Nobody really knows where moons came from, and how they came to orbit planets. Perhaps they were formed from a bit of the planet they orbit, or maybe moons were captured by the pull of the planet's gravity. Moons are always smaller than the planet they orbit, though Jupiter's moon **Ganymede**, the biggest moon in the Solar System, is actually bigger than the planets Mercury and Pluto. It is also believed that life could exist on some of the larger moons, like **Europa** which has oceans beneath its crust of ice.

Looking at the planets

Firstly, *planets emit no light of their own. They are lit up by light from the Sun*. You can see the five nearest planets with your eyes: Mercury, Venus, Mars, Jupiter and Saturn. These were the only planets the Greeks knew about because Uranus, Neptune and Pluto cannot be seen without a telescope. All five of the visible planets look much like bright stars at first sight, though Venus is always much brighter than any star, and Mercury is always hard to see because it is so close to the Sun. A few hints and tips for recognising planets are given in the information for each planet which follows. One very useful tip for all planets is that *they do not twinkle as much as stars*.

Mercury

Mercury is about 58 million kilometres away from the Sun. It is the closest planet to the Sun. It travels faster than any other planet in our Solar System, nearly 50 kilometres every second, and was named after the Roman god Mercury, the fast messenger of the gods. At one time it was thought to be the smallest planet, but with the discovery of Pluto, Mercury is the second smallest. It has hardly any atmosphere. Mercury is very hard to see because it is always so near the Sun.

The surface of Mercury looks very much like the surface of the Moon as it is covered with craters, large and small. After the Solar System had formed and for some time afterwards, there were lumps of rock whirling around. Many of them battered into the surfaces of the planets making craters. On the Earth, the craters were smoothed out by wind and water but Mercury and the Moon have no atmosphere, so the craters have remained to this day. Mercury has many huge cliffs over 500 kilometres long and three kilometres high.

This planet's 'day' and 'year' are very strange. Mercury's 'day' is 59 Earth days. A year on Mercury is 88 Earth days, as it takes this long to go around the Sun once. This means that in a Mercury year there are $1\frac{1}{2}$ Mercury days! In the time it takes the Earth to make one orbit of the Sun, Mercury has made four orbits. During a day on Mercury, the temperature can be as high as 430 °C. It has no atmosphere to hold in the heat so on the side of Mercury which is having its long night, the temperature can fall to −180 °C.

Mercury.

Mercury	
Distance from Sun	58 million kilometres
Diameter	4850 kilometres
Number of moons	0
Rings	0
Length of Day (in Earth days)	59
Length of year (in Earth days)	88
Temperature	−180°C to 430°C

Venus

Venus is named after the Roman goddess of beauty and love. The planet is a beautiful sight, shining very brightly just before sunrise or just after sunset. If you see a bright star near where the Sun has set in the evening, but you can see no other stars, it's a fair bet you're looking at Venus. Sometimes Venus can be seen before sunrise too – but you'll have to be up early to see that – especially in summer.

Venus is about 108 million kilometres from the Sun. It orbits the Sun in the same direction as the other eight planets but it rotates (turns round) from east to west. The others, like the Earth, rotate from west to east. On Venus, the Sun rises in the west and sets in the east. The planets Pluto and Uranus also have a peculiar spin (see later).

Venus is the second planet closest to the Sun and is almost the same size as the Earth. It is almost impossible to see the surface of Venus because it is covered by thick clouds. These clouds reflect a lot of sunlight making Venus shine brighter than any of the stars or other planets. Because Venus is so close to the Sun, it is always quite near the Sun when viewed from Earth. This is why you can only see it either just before sunrise or just after sunset.

People imagined it must be like Earth, only a bit warmer because it is nearer to the Sun. Some imagined a lovely world beneath the clouds, others thought it was covered with swamps with dinosaurs basking in the heat. How wrong they were! The Russians sent space probes called 'Venera' to Venus but when they went down into the clouds they stopped transmitting. Everyone was baffled. What was going wrong?

The spacecraft were crushed by the enormous pressure of the atmosphere, more than 90 times the pressure of the Earth's atmosphere. It is also tremendously hot – much hotter than the Earth and hotter even than Mercury, which is nearer the Sun. Why is it so hot? The answer is the **greenhouse effect**. Light carries energy from the Sun through the clouds and down to the surface. The light energy turns into heat, but this heat cannot pass back out through the clouds. It's as if the clouds are a blanket over the planet. This trapped heat makes the planet very hot. The same happens in greenhouses and inside cars on sunny days, except it is the glass that traps the heat.

Venus.

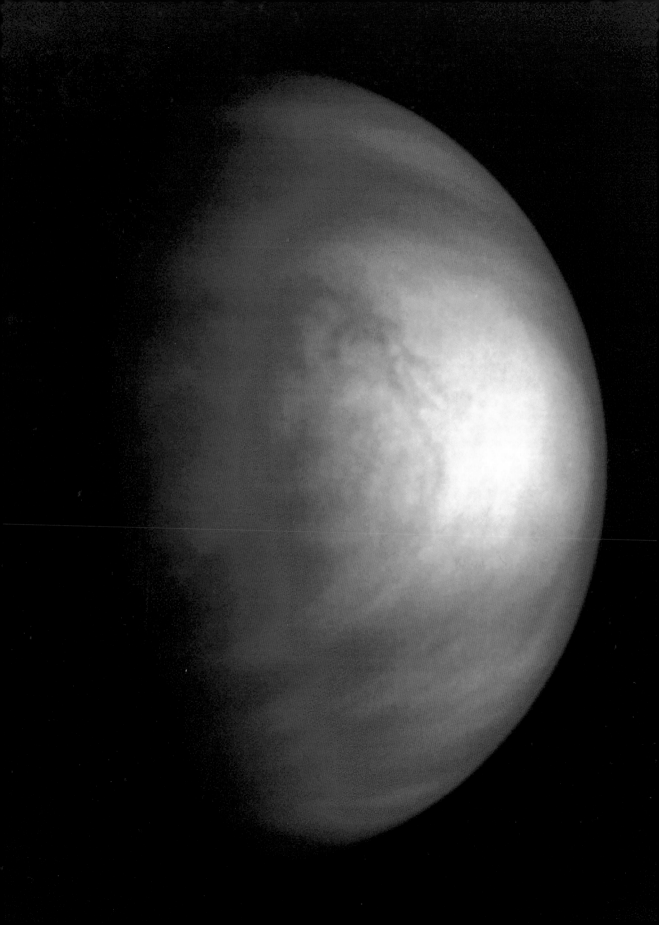

Probes, using radar to bounce signals off the planet's surface, have built up a picture of what it is like. It has mountains, valleys and lava plains but no seas – any water was boiled away a long time ago by the scorching heat. Its atmosphere is mainly a gas called **carbon dioxide**, which is what humans breathe out all the time. You would choke to death if you were breathing it in.

The yellow clouds contain acids which fall like rain. This acid rain would burn through skin and bone, so Venus is certainly not the paradise that people imagined. In fact it is a planet to be avoided unless you want to be crushed by pressure, choked by the atmosphere, burned by the acids and fried by the intense heat – all at the same time!

Venus	
Distance from Sun	108 million kilometres
Diameter	12 100 kilometres
Number of moons	0
Rings	0
Length of Day (in Earth days)	243 (spins backwards)
Length of year (in Earth days)	225
Temperature	480°C

The surface of Venus. This isn't a normal photograph, but a computer reconstruction using radar data from NASA's Magellan spacecraft which was in orbit around Venus.

Earth – a living planet

The Earth is the third planet and is about 150 million kilometres away from the Sun. It has one natural satellite called the Moon. It is a rocky planet with large oceans of liquid water. It has an atmosphere made up of different gases, roughly 75% **nitrogen**, 25% **oxygen**, with water vapour and small amounts of other gases including carbon dioxide. The Earth is the largest of the four inner or terrestrial planets.

The Earth is a very special planet – the only planet we know of which has living creatures on it. How did this happen? Many things helped to make the Earth different from the other planets.

Tolerable temperature

Venus is too hot and Mars is too cold. The Earth, between these two, is just right. Although some parts of the Earth can be extremely hot and some parts extremely cold, on the whole it is just right for life to survive.

Amazing atmosphere

The Earth's gravity is strong enough to hold on to its atmosphere. If the Earth's gravity was much weaker, the gases would have escaped into space. This is why the Moon and Mercury have lost their atmospheres. The Earth's atmosphere lets the Sun's light and warmth through but blocks most of the harmful rays of the Sun. If the Earth had no atmosphere it would have no life as we know it. Human beings and animals need oxygen to breathe and plants need carbon dioxide.

Wonderful water

If you were on the Moon looking at the Earth it would be a beautiful sight – a blue and white world with patches of brown and green. Perhaps the Earth should have been called 'Water' since more than two-thirds of its surface is covered with water. Without this water, the plants and animals of the Earth, including human beings, would not exist. Water is essential for life. Your body is made up of 20% solid matter and 80% water.

Roving rocks

Our Earth is constantly changing. Some changes happen fast, for instance a landslide or a flash-flood. Some changes are slow – a river carving out a new channel or a mountain being ground down by the wind and rain. Many of these changes are so slow that it is impossible to see them in one lifetime. Over millions of years, the Earth's surface changes. It is happening now though you are not aware of it.

Nowadays, most of the continents of the Earth are far apart, separated from each other by seas and oceans. Early in the Earth's history the continents were joined together and this allowed different plants and animals to spread over the Earth.

200 million years ago all the Earth's continents were all joined together in one big continent called Pangea. Over time the continents slowly drifted apart so that now they are separated by oceans and seas.

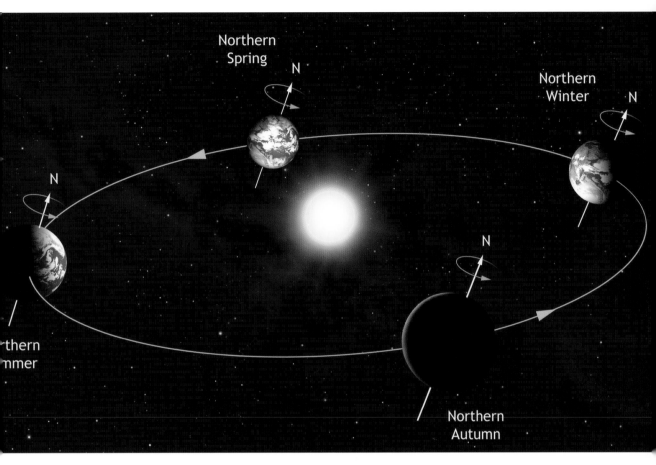

The Earth's orbit around the Sun showing its position during the four seasons.

Earth – a moving world

Day and night

The Earth is spinning round all the time. It spins around on an imaginary line called its **axis** which goes through the North Pole and the South Pole. Compared to the orbit of the Earth, this axis is not completely upright.

We all live on the surface of the Earth. When the part of the Earth where you are is facing the Sun, it is daytime for you. As the Earth spins on its axis, the part where you are faces away from the Sun and it is night time for you. The Earth spins round on its axis once every 24 hours, so 1 Earth day is 24 hours.

The seasons

As well as spinning round on its own axis, the Earth also travels in its orbit round the Sun. It takes 365¼ days to get back to its starting point, so one Earth year is 365 days. The quarter days are added up and every fourth year has an extra day – a Leap Year.

As it journeys round the Sun, the tilt of the Earth's axis gives us the four different seasons. The northern half of the Earth (the Northern Hemisphere) has summer when the North Pole is tilted in the Sun's direction. It has winter when the North Pole is tilted away from the Sun.

The southern half (the Southern Hemisphere) has summer when the South Pole is tilted in the Sun's direction. It has winter when the South Pole is tilted away from the Sun.

So wintertime in the Northern Hemisphere is summertime in the Southern Hemisphere.

Far-travelled passengers

You have probably been in a car or bus travelling on a fast road. You probably travelled at about 60 miles per hour, which is about 100 kilometres per hour. That's not at all fast compared with the Earth in the Solar System. The Earth spins around all the time on its axis at a speed of about 1600 kilometres per hour at the Equator. Away from the Equator, the speed is less but we're all moving with the Earth as it spins.

While the Earth is spinning on its axis, it is also whizzing through space on its yearly orbit of the Sun at a speed of 108 000 kilometres per hour. This is over 2.5 million kilometres per day for the Earth and all its passengers! Why are we not pushed off into space? Say a big 'thank you' to the Earth's gravity, which holds us down on to the surface while the Earth zooms around.

Far-travelled hangers-on

The Moon is the Earth's natural satellite. It orbits the Earth, held in the grip of the Earth's gravity. The Moon's own gravity is much weaker than the Earth's although it does affect the Earth in some ways (the tides). As well as this natural satellite, the Earth now has many artificial satellites, put in orbit around the Earth by people. They are placed there for many reasons: to observe weather, to help with navigation, for military reasons, for TV broadcasts and for looking out into space.

The Earth as seen from space as the astronauts of Apollo 17 sped towards the Moon.

Satellites have made communication between different continents much easier and faster. The Space Shuttle is a satellite during the time it spends in space, as was the Russian Space Station Mir. Another is the Hubble Space Telescope (HST). It gives us very clear pictures of stars, space, planets and galaxies as it orbits above the Earth's atmosphere.

Earth	
Distance from Sun	150 million kilometres
Diameter	12 800 kilometres
Number of moons	1
Rings	0
Length of Day (in hours)	24
Length of year (in Earth days)	365¼
Temperature	−80°C to 55°C

The moon

The Moon orbits the Earth, held there by gravity. It does not count as a planet because it orbits the Earth, not the Sun. It is a natural satellite (that is, it wasn't put there by us) of the Earth. The Moon is very different from the Earth. The Earth's atmosphere slows down approaching lumps of rock and usually burns them up (by friction) before they reach the ground. The Moon does not have an atmosphere so its surface is a desert of **craters**, mountains, rocks and dust.

The whole disc of the Moon as seen by the Apollo 17 astronauts.

The Earth as seen by Apollo 17 astronauts from the Moon.

Since it has no atmosphere, the sky always looks black. Anyone going there, like the U.S. astronauts, needs to take all their own air, water and food. They also need to wear a spacesuit to protect them from harmful rays from the Sun and from extreme heat or cold.

The Moon's seas

When **telescopes** were first pointed at the Moon, the flat smooth areas were thought to be seas so they were called 'mare' (pronounced 'mar-ay'), which is Latin for sea. However these 'maria' (plural of mare) or seas were found to be beds of ancient, dried-up lava which flowed a long time ago on the Moon. Mare Imbrium, the Sea of Showers, is the largest mare we can see and is about 1000 kilometres wide. When you look at the Moon, its mountains look lighter and its maria, or flat plains, look darker.

The Moon always has the same side turned to the Earth. The Moon spins round in about 28 days, but it also orbits the Earth about every 28 days. It spins just enough to keep the same face always turned our way because it is in the grip of the Earth's gravity.

The pull of gravity is not all one-sided. The Moon's gravity, although weaker than the Earth's, does affect the Earth. Wherever the Moon is in its orbit, its gravity pulls on the Earth's seas and oceans and makes the waters rise and fall, causing tides.

Phases of the Moon

You may have noticed that the Moon seems to change its shape. The Moon has no light of its own. It is lit up by the Sun's light. The same side of the Moon always faces the Earth. When the Sun shines on all of this side, we see a **Full Moon**. This happens when the Earth is almost in between the Moon and the Sun. A week after **full Moon**, when the Sun lights up only half of the Earth-facing side, we see a Moon that looks like half a circle. This is called **Last quarter** and happens when the Sun, Moon and the Earth form an 'L' shape, with the Earth at the corner. Two weeks after that, at **New Moon**, the side of the Moon facing the Earth is not lit by the Sun, and we cannot see it at all. This happens when the Moon is almost in between the Earth and the Sun. In another week, the Moon is semi-circular again, at **First quarter** and then a week after that it is back to **full Moon** again. This whole cycle takes four weeks, which is a **lunar month**, and at 28 days is a few days less than our calendar months, except February.

Each shape of the Moon is called a phase. We have described the main phases of the Moon above, but the full list, starting at New Moon is:

Day of Lunar Month	Name of Phase
0	New Moon
1–6	Waxing crescent
7	First quarter
8–13	Waxing gibbous
14	Full Moon
15–20	Waning gibbous
21	Last quarter
22–27	Waning crescent
28	Next New Moon

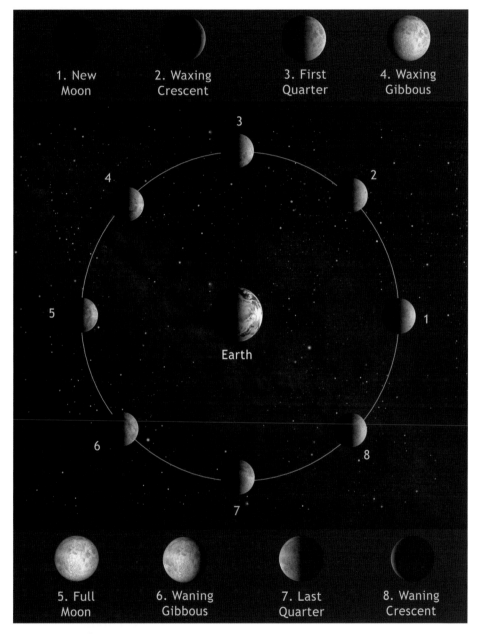

1. New Moon
2. Waxing Crescent
3. First Quarter
4. Waxing Gibbous

3

4

2

5

1

Earth

6

8

7

5. Full Moon
6. Waning Gibbous
7. Last Quarter
8. Waning Crescent

The phases of the Moon.

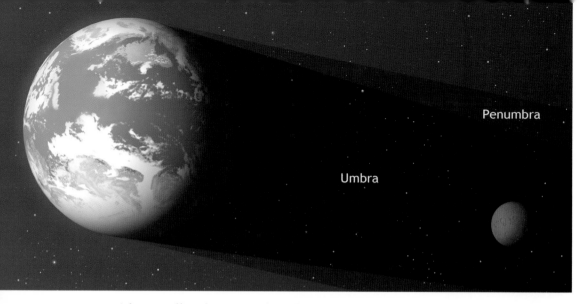

A lunar eclipse happens when the Earth moves between the Moon and the Sun.

Lunar eclipse

A **lunar eclipse** happens when the Earth casts a shadow on the Moon, blocking out light from the Sun. ('Luna' is the Latin word for 'Moon'.) During a lunar eclipse the full Moon disappears bit by bit – as if it is being eaten away – until it is all dark. Even when it is completely eclipsed some red light still lights up the Moon which makes for a very beautiful sight. Lunar eclipses are much more common and usually last for much longer than solar eclipses because the Earth's shadow is much larger than the Moon's. The Moon glows with a reddish light because it is lit by red sunlight which is bent around the Earth by the atmosphere. The blue sunlight doesn't get to the Moon because it is scattered by the atmosphere – this is why our skies are blue. You might wonder why we don't get lunar eclipses once a month at full Moon – when the Earth is in between the Moon and the Sun. The reason is that the Earth, Moon and Sun rarely line up exactly – this is the same reason why solar eclipses are rare.

Astronauts on the Moon

The Moon is the only place in the Universe, other than the Earth, that has been visited by human beings. In 1969 astronauts travelled to the Moon for the first time. This mission, called Apollo 11, was just one of the National Aeronautics and Space Administration (NASA) Apollo missions. Apollo 12, 14, 15, 16 and 17 also took astronauts to the Moon (and Apollo 13 was meant to but there was a problem on the way and the astronauts were lucky to make it home). Because the Moon has no atmosphere, all the footprints and anything else left there should remain unchanged for a very long time.

A Giant Leap

It cost a huge amount of money to put humans on the Moon. So why do it? Why were the astronauts prepared to risk their lives doing something where so many things could have gone wrong? Perhaps the main reason is that it is in our nature to ask questions and to want to explore new places.

Another reason was to prove that the theories and calculations about space flight would really work. You could say that landing a spacecraft and people on the Moon proved that it is possible for human beings to travel to the planets. The Moon is like a stepping stone to the planets. Neil Armstrong, the first person ever to step on the Moon, may well have been thinking about this when he said,

'That's one small step for man, one giant leap for mankind.'

The moon	
Distance from Earth	384 thousand kilometres
Diameter	3 500 kilometres
Rings	0
Temperature	$-155\,°C$ to $105\,°C$

Buzz Aldrin, the second human to set foot on the Moon. The photograph was taken by the first person on the Moon, Neil Armstrong.

Mars

Mars is the easiest planet to recognise because of its orange-red colour. This colour reminded ancient people of blood and war, so it was named after Mars, the Roman god of war. The red colour comes from iron in its rocks that have reacted with the gas oxygen – this makes rust, which as you know, is orange-red in colour. If you stood on Mars, you'd see its sky looking pink because of wind-blown iron dust. It is a cold desert planet with a very thin atmosphere of carbon dioxide.

Mars is the fourth planet, next out after the Earth, and is at a distance of 228 million kilometres from the Sun. On Mars the length of its day is similar to the Earth's but its year is nearly twice as long. It is about half the size of the Earth but it has a mountain three times the height of Mount Everest (the highest on the Earth) called Olympus Mons. It is the largest known volcano on any planet of the Solar System. Mars also has some huge canyons up to seven kilometres deep and stretching for 4000 kilometres. From time to time spectacular dust storms rage over most of the planet at speeds of up to 500 kilometres per hour.

Mars as seen from NASA's Mars Global Surveyor spacecraft.

A panoramic photograph of the surface of Mars taken by a camera on the Mars Pathfinder lander.

At one time it was thought that Mars had intelligent life – Martians – but the Mariner and Viking landers which landed on the surface of Mars in the 1970s found no life at all. Mars has ice caps at its North Pole and South Pole. It is now a cold world, but the Mariner 9 Spacecraft sent back clear pictures of dried-up river beds. This shows that there used to be water on Mars. It may even have had seas. There may still be microscopic life on Mars in the soil but no one is certain of this.

Mars has two tiny moons each less than 30 kilometres in size and shaped irregularly. They have been described as looking like two potatoes! One is called **Phobos** meaning 'fear' and the other is called **Deimos** meaning 'terror' – fitting companions to the god of war. They seem more like asteroids than moons and may have been 'captured' by the gravity of Mars.

The recent NASA Pathfinder mission to Mars has supplied scientists with lots of new data about the planet's rocks and its atmosphere.

Mars	
Distance from Sun	228 million kilometres
Diameter	6800 kilometres
Number of moons	2 (very small)
Rings	0
Length of day	24h 41minutes
Length of year (in Earth years and days)	1 year and 322 days
Temperature	$-111\,°C$ to $-12\,°C$

The asteroid belt

In orbit around the Sun, between the four rocky inner planets and the outer planets, there is a wide belt of asteroids, sometimes called minor planets. They are bits of rock left over from the time when the Solar System was formed. They are usually irregular in shape and come in all sizes from a few metres to hundreds of kilometres across. There are about one million pieces of rock in the Asteroid Belt but it is estimated that there are many more throughout the Solar System.

An asteroid crashing into the Earth could cause enormous damage. It is likely that any which might come close to us will be seen many years before they reach the Earth and steps could be taken to deal with them safely.

The biggest asteroid yet discovered has been named **Ceres**. It has a diameter of about 900 kilometres. Recently, it was discovered that asteroids move around in pairs, or even small groups, holding on to each other by gravity.

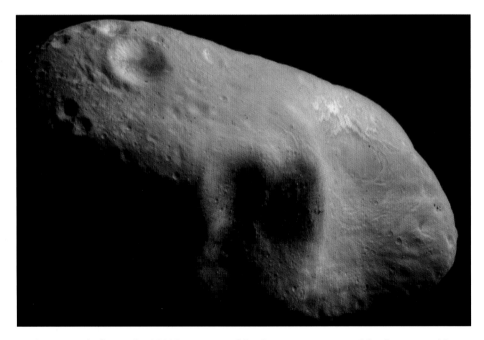

A photograph from the NEAR spacecraft's close encounter with the asteroid Eros.

Jupiter

This planet is named after the Roman god Jupiter, ruler of the gods and king of the Universe. Jupiter is the fifth planet and is about 778 million kilometres from the Sun. It is the biggest of all the planets and makes the Earth look tiny. Like the other Jovian planets, Jupiter has no solid surface. It is made up mostly of hydrogen gas with some helium. Its atmosphere has a layer of clouds 200 kilometres thick. The inside of Jupiter is very hot but its clouds are extremely cold.

If you know where to look, it is possible to see Jupiter shining brightly in the night sky. It will be brighter than all of the stars near it. With a telescope you may see Jupiter's four largest moons, its different coloured bands of clouds and maybe even the **Great Red Spot**, Jupiter's most famous surface feature. This Great Red Spot is a storm which has been raging for over 300 years. Its size, colour and brightness have varied over this time but it is still there.

Despite its huge size, Jupiter spins round very quickly. The great bands of clouds are caused by Jupiter's very fast spin. In fact, it spins so fast that its equator bulges and its poles have flattened. The cloud bands change very quickly.

Jupiter has a faint ring system. We know of 16 moons which orbit the planet, but there may be more. The four largest were discovered by Galileo in 1610.

Jupiter and its four largest moons. From top to bottom: Io, Europa, Ganymede and Callisto.

Jupiter	
Distance from Sun	778 million kilometres
Diameter	142 800 kilometres
Number of moons	16 (perhaps more)
Rings	few
Length of day	9 hours 50 minutes
Length of year (in Earth years and days)	11 years 314 days
Temperature	−150 °C

A Hubble Space Telescope view of Jupiter. On this photograph you can see Jupiter's ever-present Great Red Spot and three whitish spots which are shorter-lived storms.

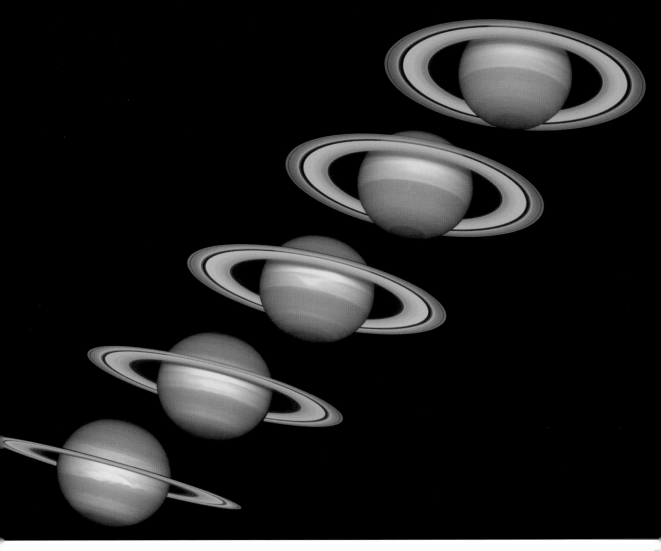

Hubble Space Telescope images of Saturn taken between 1996 and 2000. As Saturn moves around the Sun we see a different view of the planet and its rings.

Saturn

Saturn is named after the father of Jupiter. It is the sixth planet, at a distance of 1426 million kilometres from the Sun. It is the second biggest of the Jovian planets and must be the most beautiful planet in the Solar System – after the Earth!

It is surrounded by a flattened disc of rings. These rings look solid but they are made up of millions of icy rocks and snowballs. They are held in place by Saturn's gravity. The ring system is spectacular when seen through a telescope, but it can't be seen with the naked eye. Saturn is the dimmest of all the visible planets, but it is still brighter than most of the stars.

Saturn itself is made up of hydrogen and helium gases so it has no solid surface. It is colder than Jupiter, being further away from the Sun. Its cloud layers are thicker. Beneath the clouds, winds sweep around the planet at speeds of over 1600 kilometres per hour. It spins very fast and has wind and storm bands like Jupiter but they are harder to see because of a hazy layer above its clouds. Saturn is less dense than water. If you could imagine an ocean of water big enough to hold Saturn, the planet would float!

Saturn	
Distance from Sun	1426 million kilometres
Diameter	120 000 kilometres
Number of moons	18 (perhaps more)
Rings	Many
Length of day	10 hours 14 minutes
Length of year (in Earth years and days)	29 years 168 days
Temperature	−180 °C

Uranus

This planet is named after the Greek god Uranus, the father of Saturn. It is the seventh planet at a distance of about 2870 million kilometres from the Sun. Its green colour comes from ice in its atmosphere. It is believed to have a small rocky core surrounded mostly by hydrogen and helium. Its surface is a sea of liquid gases. Compared to the other planets it is very peculiar. Not only does it spin in the opposite direction, like Venus, but it is tipped over on its side. As it orbits the Sun first one pole, then the other points to the Sun. This means that many days go by on large parts of Uranus without the Sun ever rising or setting. This happens near the North and South poles of the Earth, but only for a few months at a time. On Uranus, day or night can last for 40 Earth years!

Uranus is so far away from the Earth that it was not discovered until 1781, when William Herschel first saw it using a telescope he had built himself. He had been a musician but became interested in astronomy. He wanted to call this new planet after King George, but it was decided to choose a Greek god's name since the other planets had similar names.

Uranus has a system of rings, upended like the planet itself. They are composed of stones and fine dust. These rings are much darker than the rings of Saturn or Jupiter and were only discovered in 1977. Voyager 2 sent back information about Uranus and discovered ten new moons, making 15 altogether.

Uranus	
Distance from Sun	2870 million kilometres
Diameter	51 200 kilometres
Number of moons	15 (perhaps more)
Rings	Several, but they are dim
Length of day	18 hours 10 minutes
Length of year (in Earth years and days)	84 years 6 days
Temperature	−200°C

A photograph of the planet Uranus taken by the Hubble Space Telescope.

Neptune

Neptune is named after the Roman god of the sea. It is the eighth planet of the Solar System at a distance of about 4500 million kilometres from the Sun. Sometimes Pluto's orbit takes it closer to the Sun and then Neptune becomes the most distant planet for a time.

Neptune was discovered in 1846. Two scientists, John Adams and Urbain Le Verrier, thought that Uranus was behaving strangely, as if another planet's gravity was affecting it. It was not following the path it should. They predicted that another planet would be found further out than Uranus. Astronomers started to look for this new planet and found it in 1846. So, Adams' and Le Verrier's prediction was proved to be correct – this was the first time a planet had been discovered in this way.

In 1989 Voyager 2 found that, like the other three Jovian planets, Neptune has winds and storms. Neptune's winds are the fastest in our Solar System at 2000 kilometres per hour. The winds don't travel in the same direction as the planet's spin.

Neptune as seen by the Voyager 2 spacecraft.

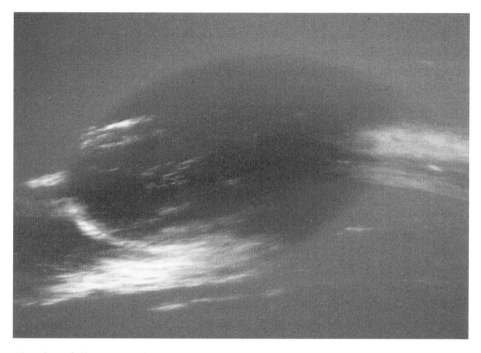

Clouds swirling around Neptune's dark spot.

Voyager 2 discovered faint rings and the Great Dark Spot. It is similar to Jupiter's Great Red Spot and big enough for the Earth to fit in it. Voyager 2 also discovered white clouds, called scooters, that fly around the planet faster than the Great Dark Spot.

Like Uranus, Neptune is made mainly of hydrogen and helium. It has an iron core and looks a beautiful blue green colour with wispy white clouds.

Neptune	
Distance from Sun	4500 million kilometres
Diameter	48 600 kilometres
Number of moons	8 (perhaps more)
Rings	few
Length of day (in Earth days)	18 hours 26 minutes
Length of year (in Earth years and days)	164 years and 288 days
Temperature	−220 °C

Pluto

Pluto is named after the god of the underworld. Usually Pluto is the ninth planet and furthest from the Sun, but for part of its orbit, it crosses Neptune's orbit and is then nearer the Sun than Neptune. Pluto's orbit is a very pronounced ellipse. Its average distance from the Sun is about 5915 million kilometres and its furthest distance is 7390 million kilometres.

It is the smallest planet and is actually smaller than the Earth's Moon. It was discovered in 1930 by Clyde Tombaugh, who died in 1998. Very little is known about Pluto. No spacecraft has ever visited it. It is thought to be like a frozen snowball of rocks, gas and water and is the coldest planet in the Solar System. As well as being an ellipse, Pluto's orbit round the Sun is tilted compared to the other planets.

Pluto is only double the size of its one moon, **Charon**. They each spin backwards like Venus and Uranus.

A Hubble Space Telescope photograph of Pluto and its moon Charon. No spacecraft has yet visited this distant world and even Hubble has difficulty making out features on its surface.

Pluto	
Distance from Sun	5915 million kilometres (average)
Diameter	2320 kilometres
Number of moons	1
Rings	0
Length of day (in Earth days)	6 days 9 hours
Length of year (in Earth years)	248
Temperature	−205 °C to 165 °C (uncertain)

Smaller things

Meteors

Meteors are small pieces of rock and metal, often just the size of dust. They enter the Earth's atmosphere but they don't reach the Earth's surface. Air friction burns them up long before they reach the Earth. As they burn up, they look like shining streaks in the sky and so they are called **shooting stars** *but they are not stars at all*. A really bright meteor is called a **fireball**. At certain times of the year you may see a **meteor shower** where many meteors appear to come from one part of the sky. The shower is named after the constellation that the meteors appear to come from, for example the **Leonids** or **Perseids**. Particles of dust and small bits of rock orbit the Sun in streams. It is thought that these streams appear when **comets** break up. If the Earth's orbit passes through a stream it causes meteor showers.

Meteorites

Sometimes lumps of rock and metal are not completely burned up in the atmosphere and bits of them fall to the Earth – these are called **meteorites**. Meteorites may be left-over bits of asteroids or comets. Most of them are small but every now and then a bigger meteorite lands.

Meteor crater.

Some meteorites can be valuable. Some of them were formed during the early life of the Solar System. Scientists are very keen to have these meteorites so that they can study them for clues about the early history of the Solar System. Very occasionally, a meteorite is found which may have come originally from the Moon or from Mars. How did they manage to get here? One theory is that a comet or asteroid may have struck the surface of the Moon or Mars which could cause rocks to be thrown out into space and eventually some have fallen to the Earth.

It is very rare for anyone to be injured by a meteorite. However, there have been some spectacularly big meteorites like one that fell in Arizona. The crater it left is known as the Meteor Crater, though it should really be called Meteorite crater. It is nearly 200 metres deep and is more than a kilometre across. It is thought that it was made about 40 000 years ago by a huge iron meteorite that was about 50 metres in size, the size of a small building.

Meteoroids

This is simply the name given to bits of rock and metal while they are out in space. So either a meteoroid can burn up in the Earth's atmosphere and become a meteor, or it can fall to the Earth and become a meteorite.

Comets

The Solar System does not end with Pluto. The solar wind continues out far beyond the last planet. The Sun's gravity can still be felt much further out, but very weakly, in the **Oort Cloud** – the home of many comets. The Oort Cloud is a vast sphere containing billions of 'dirty snowballs'. Each one is made of water, ice and frozen gases mixed in with bits of rock and dust. This dirty snowball is called the nucleus of a comet. It is a lumpy and irregularly shaped object which usually measures a few kilometres across. The nucleus of **Halley's Comet** (see later) was very dirty indeed – darker than coal in fact!

From time to time a nucleus is nudged by the gravity of a passing star. It falls inwards towards the Sun. The comet cannot be seen until it comes close enough to be heated by the Sun. As the comet approaches the Sun, it starts to melt, letting off gas and releasing the dust that was frozen into the ice. All this stuff is then 'blown' away from the comet by a 'wind' that comes from the Sun – the solar wind. This is why comets have tails pointing away from the Sun. The comet's tail, or tails, can be millions of kilometres long – much larger than the 'dirty snowball' nucleus. The comet reflects the light of the Sun and glows with a pale white light. 'Comet' comes from the Greek word *kometes* which means 'long-haired', because a comet looks like a hairy star. People often talk about comets as dirty snowballs, because they are made up of ice and dust and rock. They are much bigger than snowballs. In fact, comets can be larger than Mount Everest, the highest mountain on the Earth. Even though they are so large, they can still travel at speeds of up to 150 000 kilometres per hour – that's about a thousand times faster than the average family car can do.

Some comets are large and bright enough to be seen without binoculars, like **Comet Hale-Bopp** in 1997. Many others can only be seen by telescope. The most famous comet is **Halley's Comet**, named after Edmond Halley. He noticed that impressive comets had been seen in the years 1531 and 1607 and that he had seen one in 1682. He realised that there was really only one comet coming back every 76 years or so. He then predicted the comet would return in 1758. After it did, the comet was named after him in honour of his discovery. He didn't live to see his prediction come true, having died in 1742. Halley's Comet was last here in 1986, when we sent spacecraft to examine it. It will visit us again in 2061 – how old will you be then?

Comets take different amounts of time to make their orbit round the Sun. Comet Hale-Bopp will not return for many thousands of years. Each time a comet orbits the Sun it becomes a bit smaller because it leaves bits of itself behind in its tail.

The great comet of 1997 – Hale-Bopp. Like most comets, Hale-Bopp has two tails. The white tail is made of dust and the blue tail is made of much smaller particles that are carried away with the solar wind. This picture was not taken by NASA's Hubble Space Telescope, or at a large observatory, but by hard-working pupils and teachers at a school in Denmark.

If a comet passes close enough to a planet, the planet's gravity can change the orbit of the comet. Jupiter's gravity is so strong that it can sometimes capture and destroy comets. This is what happened in 1994 to **Comet Shoemaker-Levy 9**. First of all it was torn apart. The bits and pieces made a few orbits of Jupiter before crashing into the planet.

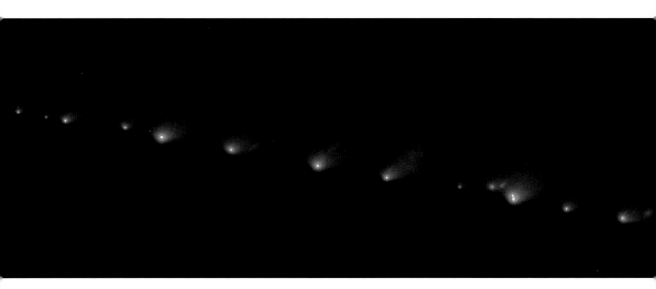

The fragments of Comet Shoemaker-Levy 9.

New comets are now named after the person or persons who first discover them. Comet Hale-Bopp was named after Alan Hale, an American astronomer, and Tom Bopp, an amateur stargazer, also an American. Carolyn Shoemaker, another American astronomer, has found over 30 comets! Carolyn, together with her husband, scientist Gene Shoemaker, and amateur astronomer David Levy discovered the Comet Shoemaker-Levy 9.

Will a comet hit the Earth?
Comets are only bright enough to be seen a few years, or even months, before coming near the Earth. Nobody can tell when one will next hit the Earth. Judging by how many comets have hit the Earth so far, astronomers think that big comets come along less often than once every million years. This means that it is unlikely that a comet will hit during your lifetime. However, there is always a very small chance that one could hit at any time.

Some astronomers believe we should be constantly looking for comets and asteroids that could be heading for the Earth. Though this seems like a good idea, not everyone agrees that spending a lot of money on setting up the telescopes is worth it. If we did find one, nobody is sure that we could do anything about it. One thing is certain – the sooner we know about one coming, the greater the chance we have of stopping it hitting the Earth. If we launch nuclear missiles at it, we could blow it up – but into how many pieces? Which is the better bet: one giant comet heading for the Earth or many smaller ones? Again, nobody knows.

Even if a comet is discovered months in advance, it's not easy to tell if it is going to hit the Earth or just fly past it. We might think a comet is going to hit the Earth when really it will miss us. If we then change the comet's path, it might hit the Earth after all – what a mistake that would be! Who could make that kind of decision?

What would happen if we were hit?

Once again, nobody really knows for certain. A big comet the size of Mount Everest would do more damage than the biggest bomb we've ever built. If it landed on a big city like London, it would easily destroy it – killing everyone there. If it landed in some remote desert, we would still be left with the problem of the dust that would be thrown into the atmosphere. This would make the sky darker, blocking out the light from the Sun, which we need to keep us alive. The Earth could spend years being a cold, unfriendly place. Most life would find it hard to survive this long winter. We know that some life can survive it because many animals and plants survived the hit that we think killed the dinosaurs.

Since most of the Earth is covered by the oceans, it is likely that the comet or asteroid would land in the sea somewhere. The damage might not be quite as bad, but still, there could be enormous tidal waves which would destroy cities and towns on the coastlines of countries.

The important thing to realise is that for the first time in human history we are able to understand, and possibly do something about, a comet or asteroid hitting the Earth. The dinosaurs lived on this Earth for a much, much longer time than humans have so far – yet they never had the intelligence or science and technology to stop themselves being destroyed.

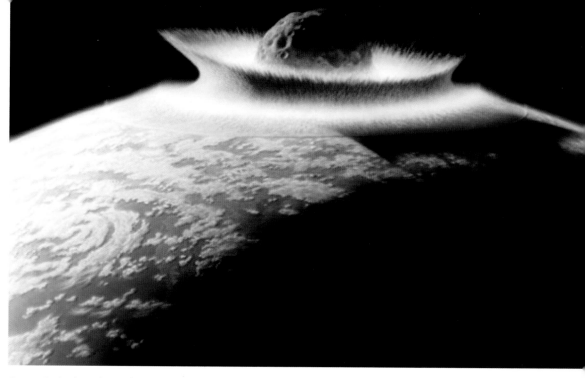

An artist's drawing of an asteroid hitting the Earth.

JUST IMAGINE!

The British astronomer, Sir Fred Hoyle, once said, 'Space isn't remote at all. It's only an hour's drive away if your car could go straight upwards.'

Let's imagine that there was a road in space and that you could travel to each of the planets in the Solar System. Sometimes a planet is closer to the Earth as it orbits the Sun and at other times the same planet is further away from us. It all depends where it is in its orbit round the Sun.

If it was possible to travel by car into space, going at a speed of 100 kilometres per hour, it would take almost 6 months to get to the Moon. To get to the Sun would take 170 years, and it's not even worth bothering to calculate how long it takes to get to the next nearest star, as it is much, much further away. That's why, when talk about everything outside the Solar System – the other stars – we have to measure distances in light-years. One light-year is the distance that light travels in one year. Light travels faster than anything else in the Universe, covering 300 000 kilometres every second. Want to know how big a light-year is in kilometres? It won't do you any good – well, here goes anyway:

1 light-year is just under ten thousand billion kilometres

or

10 000 000 000 000 kilometres

How long will the Solar System last?

As far as we know, everything in the Solar System will be here for as long as time itself. How long will that be? No one knows. But things will change. The Sun will eventually run out of energy. Just before it does, it will swell up to be an enormous red coloured star – a **red giant**. The Sun could reach out as far as the Earth, so that the Earth ends up inside the Sun for a while. When the Sun finally dies out, it will shrink down to the size of the Earth and cool down forever until one day, it will no longer give out any light. But don't worry, the Sun should last for about another 5 billion years, and if the human race survives that long, human beings may well have gone to live all over the known Universe – and maybe some of the unknown Universe!

Stars and Galaxies

Dots of light

Take a look up at the clear night sky. Every star you see there is a dot of light. You could look at any star through the best telescope in the world and it would still look like a little dot of light. But they are not dots. Each one is an enormous bright ball, full of moving and swirling hot gas. In many ways each one is like our Sun, which only looks so bright because it is very much closer to us than all the other stars.

Stars are not all the same. Some are big and some are small. Some are cooler and some are hotter. Some are very bright and very far away, some are closer but dimmer than our Sun. Some are young and some are old. Some are by themselves, but many are in pairs or in groups. How do we know all this, if stars seem like tiny points of light, even to our best telescopes? How can we know so much just from a tiny point of light? Read on.

Look, but you can't touch

Astronomy is a science, but it is different from other sciences because astronomers can really only observe. In other sciences, like chemistry, you can take two chemicals and mix them together in a laboratory and see what happens. If you made a mistake, or weren't sure what happened you can do it again, perhaps changing it or changing the way you looked at it. If a chemist finds that something really interesting happens when two particular chemicals are mixed – for example, they explode – then he or

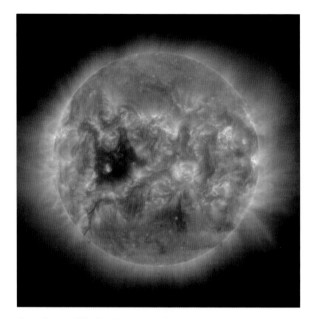

Stars look like tiny dots of light, but up close they can be much more interesting. This is a picture of our nearest star, the Sun, taken by the SoHO satellite.

she can write down exactly what they did and others can repeat the experiment in another laboratory. In this way, chemists around the world can check their results and be sure of what they have seen. They can then make theories and do more experiments to see if their theories are correct.

Astronomers cannot do experiments, they can only observe. Also astronomers can't always repeat observations. If one astronomer sees a star explode they must tell others quickly, so that they can look with other types of telescopes. They might not get a second chance.

Astronomers only see stars as tiny points of light – they can look, but they can't touch. So is astronomy all guess-work? No, as we said, it is a science, but a science that works differently from other sciences. In fact it relies on sciences done in laboratories on the Earth. Whenever a scientific theory is tried and tested in a laboratory on the Earth, we assume that it'll be true all over the Universe. For example, we know that if we drop a ball on the Earth, it will fall because of gravity. If we went to another planet, we'd expect the ball to fall there too, and that gravity works the same way everywhere.

We can't know for sure that all our ideas are true until we go everywhere in the Universe, but so far astronomy has told us that the same scientific laws are true all over the Universe – they are **universal laws**.

The big clue: star colours

You can see for yourself that stars are different colours. Look at the constellation of Orion (see *Our View of the Universe*). The star in the left shoulder – *Betelgeuse* – is red, but the star in the right foot – *Rigel* – is white, with a hint of blue.

You will have seen an electric element at some point. The bar in an electric fire is an electric element, or if you have an electric cooker, the hobs that heat the saucepans are electric elements. If you turn one on you will feel a burst of heat from it even before it glows. Then it will start to glow red, and then, as it heats up, the colour will turn paler, and become a yellow-orange colour. If you could somehow increase the electricity flowing through it, it would eventually turn white hot and melt.

What has this got to do with stars? Well, you can see from a common household object, like a fire or a cooker, that things change colour as they heat up. Could the colour of a star be telling us how hot it is? Could it even work the same way as for the fire, with red stars being cooler than white stars? From just the colour alone can we tell that *Betelgeuse* is not as hot as *Rigel*? We can, and what is more, it is true for the same reason. This is a basic law of science that is as true for stars as it is for an electric cooker or an electric fire: hot things give out light with a colour that depends on how hot they are.

Now you know the secret of astronomy. We don't just have a point of light to work with, we also have the colour of the light. In fact, almost all light that you see is a mixture of colours. If all the colours are present in light, then our eyes see white light. You can see this best with a prism, shown in the picture on the next page, which is a block of glass that splits light up into its different colours. Water droplets in the sky do this to sunlight, which makes a rainbow.

By using prisms and other inventions, astronomers can tell exactly how much of each different colour is in a star's light. By comparing this with the colours from heating objects in a laboratory, astronomers can tell how hot a star is – all this just from the colours in that tiny point of light.

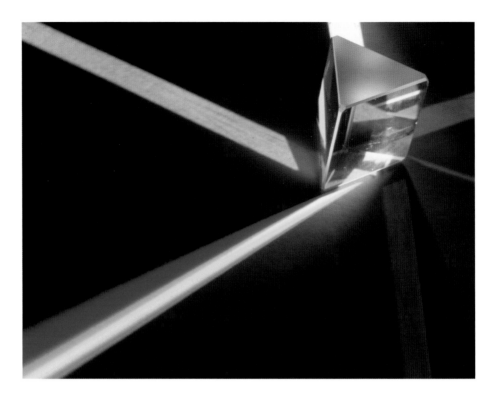

A beam of white light enters a glass prism and the light that comes out is split up into its different colours.

All stars are much hotter than the temperatures that we are used to on the Earth. A red star, like *Betelgeuse*, is about 3000 °C. Our Sun, a yellow star, is about 6000 °C. The blue-white star like *Rigel* is more than 10 000 °C.

It turns out that once you know a star's colour, you can work out many other things about it. From making observations, astronomers have discovered that most red stars are cooler and dimmer than blue stars. This is true for the stars *Betelgeuse* and *Rigel*. Also, **blue stars** tend to have more mass than the red stars. In other words there is much more gas in a blue star than a red star.

When you split light up into its different colours, the pattern you see is called the **spectrum** of the light. A rainbow shows you the spectrum of sunlight. If you were to look very closely at a spectrum – like the one in the picture on the next page – you will see that there are dark lines. These are called **spectrum lines**.

Normally you can't see spectrum lines with your eye, but astronomers have built special devices called **spectrometers** that can look at these lines closely. A collection of spectrum lines is like a finger-print that tells you that a particular gas is present. By comparing the spectrum of a star with the spectrum from a gas in a lab, astronomers can see which 'fingerprints' are present in the light. After a bit of detective work, this can be turned into a list of different gases. In fact, by looking at how dark the lines are, you can tell how much of each gas is there and even the temperature of the gas. It turns out that normal stars are nearly three-quarters hydrogen and one-quarter helium. The word 'nearly' is used, because stars do contain a small amount of other gases, whose amounts do vary a lot from star to star.

Snapshots of the stars

We can tell so much about stars by their colour but, as we said before, that's not the whole story. Stars are born, grow older and will eventually die. As they grow older, they can change. For most of a star's life it stays the same. The Sun has changed little over the last four billion years (4 000 000 000 years) and will not change much in the next four billion years. Astronomers have a pretty good idea of how stars change over their life-times, and they've worked all this out in the last 100 years. If stars can take so long to change, how could astronomers have worked it out so quickly? Stars can live for billions of years, so 100 years is quick by star standards.

The spectrum of the Sun. The lines are caused where gases in the atmosphere of the Sun block out certain colours of light.

You might start by wondering if small, dim, red, cool stars grow to become big, bright, blue, hot stars. Imagine you are an alien biologist, travelling from one planet to another, studying any life that you find. You don't have long to spend on each planet, so all you can do is take photographs – snapshots – of what you see there. Imagine you land on the Earth, which is full of life. If you spent only a day on the Earth you couldn't watch humans, or most other forms of life get older. So, if you weren't a very good alien biologist you might make the following theory.

Ants grow to become mice, which turn into dogs, which turn into humans which will eventually grow up to become elephants, which eventually get too big to walk so take to the sea to become whales.

Obviously, anyone who has ever lived on the Earth for any length of time knows that this isn't true. Old people do not turn into elephants! In the same way, someone who has studied science on the Earth would be able to show from universal laws that small, dim, red, cool stars do not grow to become big, bright, blue, hot stars. They are different because they were born different. Red stars start out with less mass than blue stars. In fact, by knowing the basic laws and how much mass a star starts with, astronomers can see what will happen to stars in computer simulations. These simulations tell us that the Sun should live for about ten billion years. This makes sense, because astronomers believe that the whole Universe is older than this, and geologists can tell us that the Earth and the Sun are about four and a half billion years old.

But, we still haven't answered the main question. How do we know these computer simulations are right about stars, when we haven't had time to watch a star grow older? The answer is that when we look at the night sky, we see a snapshot of stars at different points in their lives. If the alien biologist visited a city, there would be a whole range of ages on show. By looking at the size of person, their skin and their hair and the way they moved, you could guess their age reasonably well. In the same way, from our knowledge of science on the Earth and from using computers, we can tell a lot from just looking at the different types of stars that we see.

The typical person will be at school or college up to the age of about 20, from ages 20 to 70 they will be 'working' and after 70 they will be 'retired'. So, many people spend 20 years 'learning', 50 years 'working' and probably ten years being 'retired'. If you were to do a survey of people living in a town or city, you would find they are mostly 'working'. This is because we spend most of our lives working.

Astronomers do something similar with the stars. From their computer simulations they have good estimates of how long different types of star will live for. Stars with the most mass, that is stars with the most gas, live and die in a much shorter time than stars with less mass. The reason is that heavy stars use up their 'fuel' faster. Massive stars burn hotter, brighter and faster than less massive stars, that are cooler, dimmer and live for longer. So a star with ten times the mass of our Sun will be 10 000 times brighter, but will die in ten million years, whereas our Sun's lifetime is about ten billion years. If this theory is correct, then massive stars should be rarer than less-massive stars. By looking at the numbers of the different kinds of stars in the sky, astronomers can show that this is true and that it fits in with the computer simulations. In this way, with the laws of science and the help of modern computers, we can tell the life-story of the stars in the night sky, all from just a snapshot.

The star in this Hubble Space Telescope image is believed to have started out its life being 200 times the mass of our Sun. It is ten million times brighter than the Sun, and would reach all the way out to the Earth if it were in our solar system. Compared to our Sun, this star is very young, probably just a few million years old. In a few more million years astronomers think its life will end in a gigantic supernova explosion.

The life of stars

We said that stars are born, then live and eventually die. But they are not alive, we just use the words 'born', 'live' and 'die' to make stars seem more familiar. They are 'born' when a cloud of gas is pulled by gravity into a smaller and smaller ball. Imagine you are in a room full of people, and every person in the room tried to walk towards another person. Very soon everyone would be huddled in groups, and eventually everyone would get together in one big group. The gas in clouds does the same thing, but is pulled together by gravity.

Now, if the Sun were still shrinking we would see it getting smaller in the sky. But it isn't shrinking anymore – why not? Has its gravity 'turned off'? No, it can't have done, because the planets, including the Earth, would fly out into space away from the Sun. The answer is that as the gas in the Sun is squashed into a tighter and tighter ball, it pushes outwards more and more. Try squeezing a balloon (not so hard that it bursts). As you push it in, it pushes back. As soon as you stop pushing on the balloon the air in the balloon will push the balloon back into shape. So stars like our Sun remain in a ball shape for most of their lives because gravity pulls the gas inwards just as strongly as the gas pushes outwards.

As a young star shrinks it also gets hotter. Eventually the temperature in its core – the very centre of the star – gets so high that nuclear reactions start. These reactions are called **nuclear fusion**. You might have heard that we use nuclear power on the Earth – this is **nuclear fission**. No scientist on the Earth has yet managed to get nuclear fusion working in a useful way. It may be that we have to learn to re-create conditions like the inside of a star before we can get it to work. But, in a way, our most important source of energy *is* nuclear fusion already – the nuclear fusion taking place at the centre of the Sun. Without it the Sun wouldn't shine and most, if not all, life on Earth would die out. If it weren't for nuclear fusion power-ing the sunlight, we would have no food, no natural light and no weather!

Nuclear fusion gives out energy by turning hydrogen into helium. Three-quarters of the Sun's mass is hydrogen, and most of the remaining quarter is helium. As time goes on, hydrogen in the core, where nuclear fusion takes place, will gradually be turned into helium. Once most of the hydrogen in the core is gone then the life of the star is virtually over. It is hydrogen that fuels a star and once it is gone the star's life is at an end. What happens to a star then depends on whether it is big or small.

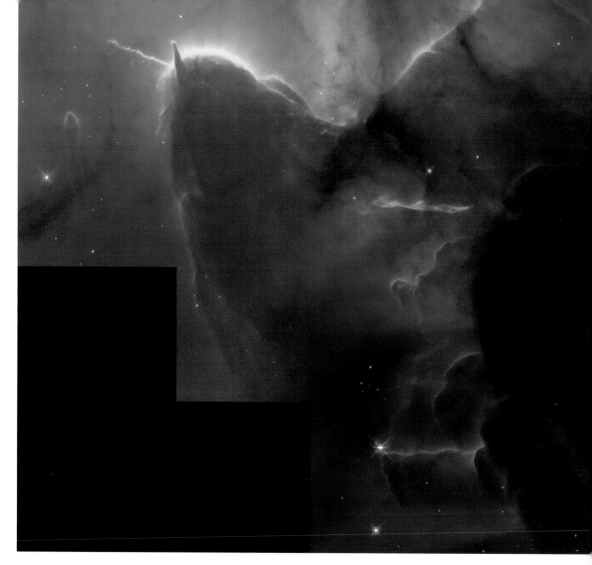

The Hubble Space Telescope shows us a stellar nursery being blown apart by light coming from a massive, young, hot star just off the top of the picture. The jets in the top left of the picture are thought to be from stars that are just forming in this harsh environment.

Big stars and small stars

The length of a star's life, and how it will die, depends on how big the star is to begin with. The words 'big' and 'small' are a bit vague. What is really important is the mass of the star. If the star is 'big' – meaning it has a *large* mass – then it contains more hydrogen fuel. Strangely enough, big stars live for much shorter times than small stars – stars with a small mass. The reason is that big stars have more gravity, and their core is pulled into a tighter, hotter ball. Because it is so hot, the nuclear reactions speed up and gobble up the fuel much faster. The big stars are brighter and hotter, live faster and die younger.

The biggest star we know of is about 100 times the mass of the Sun. Stars any bigger than this cannot form because the inside would burn so vigorously that their outside would be blown off into space. Stars less than about ten times smaller than the mass of our Sun don't have enough gravity to start the nuclear reactions. They cannot pull the gas into a tight enough, hot enough ball for fusion to start happening. These failed stars are known as **brown dwarfs**, and are like bigger versions of our planet Jupiter.

The death of stars

Not surprisingly, the big stars that live the faster, more energetic lives have the most dramatic end. Once they have exhausted their fuel, their insides will cool and the pull of gravity inwards totally overwhelms the outward push of the cooling gas. The star collapses. But, there comes a point where the gas in the core is suddenly able to halt gravity. At this point an *enormous* amount of gas is moving *very* fast inwards, towards the core which is like a solid wall. All this gas is stopped by hitting this 'wall', then it bounces off the core and out into space. During this violent time, all kinds of weird and wonderful reactions are taking place. Light of all colours – visible and invisible – pours out. This spectacular stellar death is called a **supernova**.

Wouldn't it be amazing if we could see a supernova going off? Unfortunately, a supernova is not a common occurrence. Although they are rare, they are so bright that we can see them from a very long way off. In 1987 a supernova went off quite near to us, and using modern telescopes and cameras, astronomers got a chance to study it in more detail than ever before. Have a look at the before and after picture. A faint, hardly noticeable star suddenly turns into a star that you could see with your own eye, without any kind of telescope. What's more, this star isn't even in our own galaxy (we'll come to galaxies later).

About ten years later, the Hubble Space Telescope had a closer look and saw rings around the supernova. It turns out that these rings didn't come from the supernova explosion, but are from a previous mini-explosion. The light from the supernova explosion has lit these rings up.

Before (top) and after (bottom) the 1987 supernova. The spikey shape is nothing to do with the star itself, but is a result of the star being so bright in this photograph.

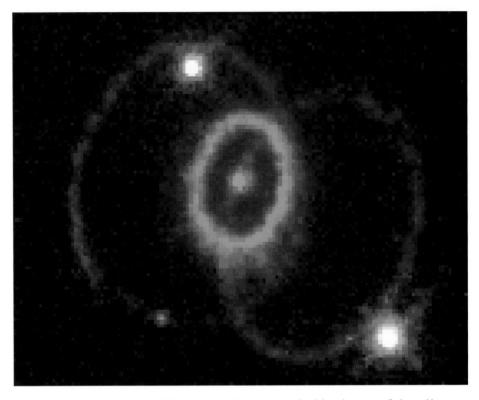

Some ten years on the Hubble Space Telescope took this picture of rings lit up by the 1987 supernova.

In 1054, Chinese astronomers saw a new star appear. This new star could be seen during the day and lit up the night so that you could read a book by its light. The Chinese astronomers didn't know it, but they had seen a supernova go off in a region of space not too far from our Solar System. Today we can still see this explosion in the patch of sky called the **Crab Nebula**, though now it is not nearly so bright and has spread out quite a bit. We'll return to the Crab Nebula later.

When a supernova explodes it can leave a small, very dense and spinning star at its centre. This is called a **neutron star**, which we'll come back to later.

The death of smaller stars is also dramatic, but in a different way. Once a star like our Sun has used up most of its nuclear fuel, it will swell up and become a **red giant**. When it becomes a red giant the Sun will become more than 100 times bigger. The red giant Sun will engulf the three inner planets (Mercury, Venus and the Earth) and reach out towards Mars.

Antares

Rigel   Sirius A   Sun

The Sun and some larger stars drawn to scale.

When the very last bit of nuclear fuel is used the outer part of the Sun will be thrown off into space. The inner part of the Sun, which was the core, will cool and gravity will pull it down into a ball about the size of the Earth. This will leave a dim star called a **white dwarf**. The brightest star in the sky – Sirius – has a white dwarf companion. The star that we can see so brightly is called Sirius A, and its small, invisible to the naked eye, companion white dwarf is called Sirius B.

Dead stars

When a human dies, all the parts of the body that kept it alive stop working. When a star dies no more energy is made in its core because nuclear fusion has stopped. Without any energy to fight against gravity, the star will shrink as the gravity from the middle pulls the layers on the outside inwards. If this shrinking by gravity carried on in this way then the star would eventually become a point, a little dot in space, but we know this doesn't always happen.

A star that is not too big, no more than double the mass of our Sun, will probably end up as a white dwarf. In a white dwarf the shrinking by gravity is stopped when it reaches the size of the Earth and the particles which make up the star are pushed very, very close together. Just a sugar-cube-sized bit of white dwarf material would weigh a tonne if you brought it to Earth. In other words, a white dwarf sugar-cube would weigh more than a car!

A star which starts off by being quite a bit bigger than double the mass of our Sun will become a supernova. After it has exploded, what is left shrinks down to a size even smaller than a white dwarf, until it is only about ten kilometres across – about the size of a city on the Earth. This is a **neutron star**. One sugar-cube-sized lump of neutron star would weigh 1 000 tonnes on the Earth, that's a whole car park full of cars!

Take a round, drinking glass and look through the side of it at the room around you. It will seem all distorted. When light passes through glass or water, its path can be bent and so what we see can seem distorted. What has this to do with neutron stars? Well, gravity can also bend light. Although there is gravity from every object around us, none of it is strong enough to bend light enough so that we would notice it. The gravity near a neutron star can noticeably bend light. If you could stand on a neutron star with a telescope, you could see someone waving at you on the other side. That's like being able to see someone in Australia waving at you!

In 1967 a signal was detected from space that gave a repeated blip in radio waves every second. It was so regular that some people thought it was LGM – Little Green Men from space! This radio source, which is called a **pulsar** is actually like a kind of natural inter-stellar lighthouse. A pulsar is a neutron star with a 'radio hotspot' – a patch on its surface that gives out very strong radio waves. Every time the neutron star turns a beam of radio waves sweeps across the Earth, giving us a blip in radio waves. Using **radio telescopes** on the Earth astronomers have found that more than a hundred pulses every second come from some pulsars. The amazing thing is that this tells us that some neutron stars are spinning hundreds of times every second. That's an object more massive than the Sun, but much smaller than Earth turning hundreds of times every second! Compare this with the Earth which takes 24 hours to spin round once, or the Sun which takes 27 days to spin round once.

Proxima Centauri

Sirius B

Earth

Sun

The Sun and the Earth and some smaller stars drawn to scale.

The picture shows some small stars compared to the Sun. Proxima Centauri is the nearest star to us, but is still very dim in the sky because it is so small and dim itself. Sirius B is a white dwarf. A neutron star would be smaller than a speck of dust on this diagram.

There are things in the Universe that we don't really understand. One such thing is the **black hole**. Scientists know how they might form and are quite sure that they exist. Stars more that twice the mass of our Sun will keep collapsing, and will end up as a black hole. If you get too close to a black hole it will suck you in and you will *never* escape. The fastest thing we know in the Universe is light. It travels 300 000 kilometres every second. Light can travel across the Earth in a blink of the eye. But even light feels the effect of gravity and if it gets too close to a black hole, it cannot escape. If light can't escape then nothing can.

Picture a ball-shaped surface around a black hole. Outside, light is free to move about the Universe. Inside, light is trapped forever. This surface of no-return is called the **event horizon**. Anything that crosses the event horizon into the black hole is lost forever, and will never escape.

That is why they are called 'black holes'. A 'hole' because things disappear inside them, 'black' because light can disappear inside them and the absence of light is 'black'.

How do we know black holes exist if they are so black? Well, their strong gravity can have a tremendous effect on what is around them. If they are near another star, then we might notice the effect of the strong gravity on the movement of the other star. Also, if it is close enough to another star, a black hole will suck gas off the outside of the star. Eventually this material will fall into the black hole, but before it does, it can get very hot and even explode. These explosions can be seen as stars suddenly becoming very bright. A suddenly brightening star is called a **nova**.

Another interesting way to see black holes is to see their effect on passing light. Usually, in astronomy, we know something is there because it gives out a lot of light. The brighter it is, the further away we can see it. Unless there is another star nearby, you might think there is no way we can see a black hole, since it is as black as empty space itself, but that is not quite true. If a black hole happens to pass in front of a star then the star can suddenly flash. This happens because the black hole bends the light and brings many more rays of star light to the Earth. It is like looking at cars driving by on a sunny day. If a car drives by with one of its glass windows at just the right angle, it can reflect the Sun for an instant and you will see a flash of light. Sometimes you can see a flash even when you can't see the car. In the same way, sometimes astronomers see a flash of star light and know that something, probably a black hole, passed in front of the star, even though they didn't actually see it.

Nebulæ: clouds in space

On the Earth, a white cloud sitting in a blue sky is a common enough sight. If you could look closely enough you would see that a cloud is just air with an enormous number of little droplets of water or ice crystals in it. In space you get clouds too. Like just about everything in the Universe, these clouds are mainly made up of hydrogen gas. Clouds often have dust grains in them and astronomers have found other surprising things in them, like water. We call such a cloud of gas a **nebula**. If we're talking about more than one we don't call them 'nebulas' but 'nebulæ', pronounced neb-yoo-lay.

The Pleiades star cluster.

We can see a nebula in a number of ways. If there is a star in a nebula it will light up the gas and dust, a bit like a light on a foggy night. A big telescope shows this happening in the **Pleiades**.

Sometimes the sort of light from a star can make the gas glow, rather like the gas in a fluorescent tube in a strip-light. You can just about see this with your own eye in the Orion Nebula, which is just below Orion's belt.

Isn't it good luck for us astronomers that these stars are there? Well, not really – it's not *just* good luck. Those stars are there because stars form in gas clouds or, in other words, in nebulæ. Astronomers have seen stars in the Orion Nebula – the Proplyds – which are just forming. Many of the other stars in the Orion Nebula are 'babies' too. Since stars can live for billions of years, a star that is a few million years old is still a 'baby' star.

The Orion Nebula.

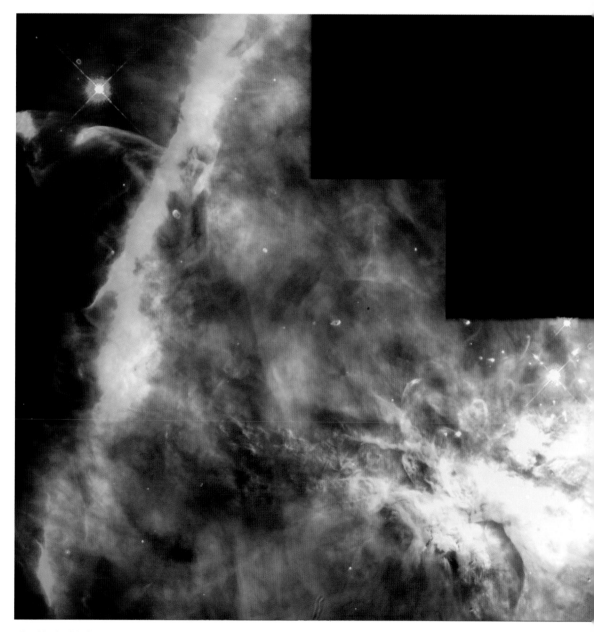

The little blobs, called Proplyds, seen in this Hubble Space Telescope image are young stellar systems that will one day become much like our Solar System.

The Horsehead Nebula.

The Pleiades is a kind of star nursery. It is a group of stars that started forming about 65 million years ago, about the time the dinosaurs died out here on the Earth. The wispy bits in the photograph are left over bits of gas in the nebula, lit up by the light of stars.

We've said that a nebula can be lit up by stars or that stars can make the nebula glow, but there is a third way to see a nebula. Have a look at this photograph. It is of the Horsehead Nebula – but what is the horse's head?

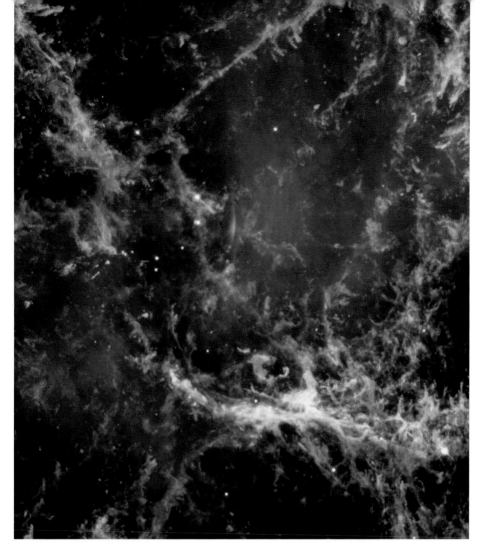

The Crab Nebula – an explosion that has being going on for almost 1000 years.

If you look closely you will see that the horse's head isn't a hole in the nebula, or a gap – you can't see any stars at all in it. In fact, it is just another cloud of gas not lit up by anything, blocking out the light from the nebula behind. It's a bit like someone standing in front of a bright light. All you can see is their shape in black – a silhouette.

As well as being the place where stars are born, a nebula can appear when a star dies. In 1054 Chinese astronomers looked up and saw a star in the sky so bright that they could have read by its light at night. This star was so bright that it could even be seen in daytime. This was a massive star that had exploded – a supernova. Today even a small telescope can show that there is a nebula in this patch of sky – the Crab Nebula. This nebula is an explosion that has been going on for almost 1000 years. At the centre of the Crab Nebula is a neutron star, left over from the star's explosion.

Star clusters

There are over 100 stars in the Pleiades, all born together from gas from the same nebula. We call this a **star cluster**. A constellation like Orion isn't a star cluster because the stars in Orion aren't close to one another and certainly weren't all born in the same region of space. In fact the Pleiades is about as big as star clusters get in the night sky. Clusters like the Pleiades have no particular shape and so are called **open clusters**. Our Sun was probably born in an open cluster, but it has long since left that region of space, and we will probably never know which stars were born in the same cluster as our Sun.

One of the most spectacular sights through even a small telescope is another kind of star cluster – the **globular cluster**. 'Globular' is pronounced glob-yoo-lar.

The globular cluster M13 – a ball of stars in space held together by gravity.

There can be millions of stars in a globular cluster. These stars can be very old. Some are almost as old as the Universe itself. All the stars in a globular cluster feel the pull of gravity from all the other stars. This is what gives the cluster its shape, with most stars being pulled into the centre of the cluster. Look at the photograph. You can barely tell one star from the next in the middle of the cluster.

Galaxies

We've talked about star clusters. A cluster of stars can have anywhere from just a few stars to many millions of stars, but these clusters are tiny compared to **galaxies**. In fact, star clusters are just small pockets of stars in galaxies. A galaxy is like an enormous star cluster. The stars in a galaxy often form a disc with a big lump in the middle. All these stars go round the lump in the middle, rather like the way the planets orbit the Sun. Astronomers believe that there is a big black hole in the centre of most galaxies, a black hole that is constantly gobbling up other stars that come too close.

A group of five galaxies known as Stephan's quintet.

The Sun and all the stars in the sky are in a galaxy called the **Milky Way**. It is nothing special as galaxies go, containing about 100 000 000 000 stars (count those zeros!). Say that number in words: one hundred thousand million, or if you want to save time you can just call it one hundred billion. Can you imagine this number? I can't, I don't know anyone who can really say, honestly, that they can picture one hundred billion of anything. It is an unimaginably large number. There are about six billion people on the Earth. If we were to share out the stars in the Milky Way amongst everyone on the Earth, then we'd get about 17 stars each.

Galaxies are very big, so big in fact that there is no point measuring their size in metres, kilometres or miles. The best way to do it is to work out how long it would take light, the fastest moving thing in the Universe, to get from one side of a galaxy to the other. It would take light about 80 000 years to get from one side of our galaxy to the other – so astronomers say that the galaxy is about 80 000 light-years across. Remember a light-year is a *distance*. It takes light just over four years to get from the Sun to the nearest star (Proxima Centauri) – so it is four light-years away. It takes about eight minutes for light to travel between the Sun and the Earth – a distance of eight light-minutes. So, you can see how big a galaxy must be compared to our Solar System. Our galaxy is surrounded by globular clusters, each one a cluster of millions of stars. The Sun is some way out from the centre of our galaxy and is on one of the spiral arms.

Have you ever seen the Milky Way? You need to be well away from streetlights to see it. You have to get out of towns and cities and find a nice dark place in the countryside. If the sky is dark enough, you can't miss it. The Milky Way is a fuzzy glowing strip of light that stretches across the sky. What you are looking at is our galaxy, but from the inside. All the millions of stars blend into each other to make it seem like a milky band of light. If you get a chance to look at the Milky Way, try to imagine what you are looking at. Try to picture yourself on the planet Earth going around the star called the Sun, which is one of several hundred billion stars in our galaxy, the Milky Way. Can you see why we would look up from inside our own disc-shaped galaxy and see a band of light stretching right across the sky? One word of warning – while you are gazing up at the sky in the dark being wowed by the wonders of the Milky Way, try not to lose your balance and fall over and hurt yourself – it has been known to happen!

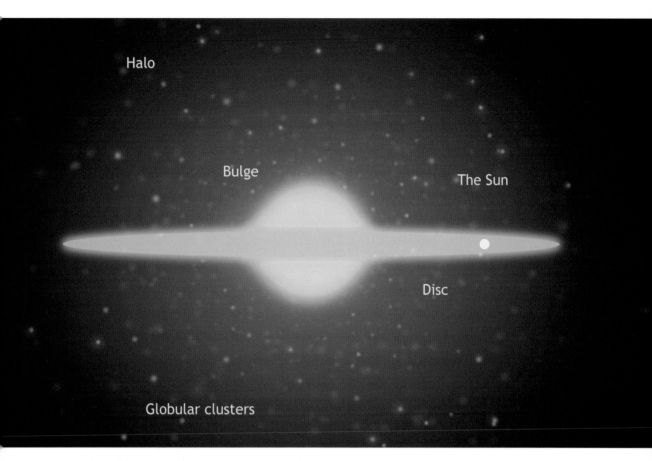

Halo

Bulge

The Sun

Disc

Globular clusters

The Milky Way galaxy – our galaxy – viewed edge on. The Milky Way is disc-shaped, with a central bulge, surrounded by many globular clusters.

If you know where to look, you can see another galaxy – the **Andromeda Galaxy** – with your eyes. Again, you need to get a reasonably dark sky, but it's actually easier to see than the Milky Way because it's not spread thinly across the sky. With the eye you just see a smudge, but with a really big telescope, and some state-of-the-art equipment, you can get an image that is literally fantastic.

The Andromeda Galaxy is our nearest neighbour, being just over two million light-years away from us. That means that the light we are getting from the Andromeda Galaxy now had to have left two million years ago. If you think about it, that means that we are seeing this galaxy as it was two million years ago.

The Milky Way and the Andromeda Galaxy are very similar. They are about the same size, have the same number of stars and both are what are called **spiral** galaxies. You can't see this when looking at the Milky Way in the sky because we are inside it. In really good pictures of the Andromeda Galaxy, you can see that it is a spiral, but it is still difficult to see because Andromeda appears side-on to us. Some galaxies,

M31 – the Andromeda Galaxy.

just by chance, are at just the right angle to show their spirals really well. One such galaxy is the **Whirlpool Galaxy**.

M51 – the Whirlpool Galaxy.

Not all galaxies are spirals. Some galaxies are like squashed rugby balls and are called **elliptical** galaxies, while others have a ragged shape and are called **irregular** galaxies. Astronomers are not sure why galaxies have the shapes they do, but sometimes we can tell that irregular galaxies have formed as a result of two galaxies colliding.

M87 – An elliptical galaxy.

The Andromeda Galaxy and several others are in what is called **the local group** of galaxies. Like stars, galaxies tend to cluster together. That is the effect gravity has on things – it pulls them together into clumps. Clusters of galaxies like our local group are common. In fact clusters of galaxies cluster together to form enormous **super-clusters**.

M82 – An irregular galaxy.

The Universe

There are least as many galaxies in the Universe as there are stars in our galaxy. The Universe is everything – all of space, every planet, every star, every galaxy, every galaxy cluster, every super-cluster – everything.

Where does the Universe end? This might be one of those questions that doesn't have an answer because the question doesn't mean anything. Think back to the Earth: what's north of the North Pole? If you go north, you are heading towards the North Pole. When you get to the North Pole you can't go any further north, so nothing is north of the North Pole. We haven't mapped out the Universe as well as the Earth, so at the moment we can't tell for sure whether it ends or not. There are two possibilities: either it never ends – it is **infinite** – or it has a certain size – it is **finite**. If it is infinite then it just goes on for ever with no end. If it is finite, it is only a certain size, and may or may not have an end. This probably seems confusing, so we'll come back down to Earth for a bit.

When people believed the Earth was flat, they were scared that if they travelled too far they would fall off the end. But now we know that the surface of the Earth has no end. If you set out travelling, you would keep going for ever, going round and round, finding no end. So, like the surface of the Earth, the Universe might have a certain size, but have no end.

One thing we do know is that every galaxy we look out on is moving away from our galaxy. Are we in a giant explosion with us at the centre? If so, that makes our galaxy special. The answer is no. Every galaxy is moving away from every other galaxy. If we were in another galaxy, we would still see all the other galaxies racing away from us. How can this be? Well, imagine putting lots of dots on a balloon that hasn't been blown up. When you blow it up, every dot will move away from every other dot. (Find a balloon, a felt-tip pen and a mirror and prove it for yourself.)

We now know that the entire Universe is expanding, like a balloon being inflated. If it is getting bigger, it makes sense that it was smaller in the past. If we go back in time far enough, the Universe must have been very small – just the size of a point. The **Big Bang** theory is along these lines. It says that the Universe started as a point about 12 billion (12 000 000 000) years ago. It exploded and then kept getting bigger until it reached the size it is today.

A Hubble deep field image. How many galaxies can you see?

We know that it will keep expanding for a while, but we are not sure if it will carry on forever, or if the Universe will stop expanding and start to shrink. If it starts shrinking, then we might end up in a **Big Crunch** as everything comes back down to a point. A sort of Big Bang in reverse. If it keeps expanding, then eventually the stars will use up all their fuel and the Universe will become a dark waste-ground, littered with small rocky planets, cool blobs of dust, dim white dwarfs, neutron stars and black holes. At the moment astronomers think it will keep going – but don't worry, the Universe won't change much for long time.

The study of the entire Universe is called **cosmology**, and the Big Bang is the theory that gives us the best explanation of the facts we've discovered with different kinds of telescopes. As we build new telescopes and make new discoveries we could well find that the Big Bang theory is either wrong or just isn't the whole story.

Life elsewhere in the Universe

We know of only one place in the entire Universe where there is life – here on the Earth. But we know of other places in the Universe where life like we have here could exist. One place is Mars, or at least Mars millions of years ago. Since we can now go to Mars, we can look for evidence that life is there and look for fossils to show that life was once there. So far (in the year 2001) we have only sent space-probes. Some of them tested for life in a particular way, but found nothing. Scientists have also found some evidence of life on a meteorite found on Antarctica. This meteorite came from Mars, but we can't be sure that's where the life came from. It could just be life from the Earth that got on to the meteorite after it fell. We've only just scratched the surface in our search for what is called **Extra-Terrestrial** (ET) life, and we have lots more tests for different kinds of life, dead or alive.

If life exists in other places in our Solar System, then it is bound to be quite a bit different from life on the Earth. Since we've only ever had a chance to examine life on the Earth, we can't even make guesses as to what other kinds of life could exist in other places where we could never live.

We know of many stars like the Sun, which could have planets like the Earth circling them. We have also seen that other stars have planets too. At the moment we can only see big planets – planets like Jupiter,

Saturn, Uranus and Neptune. Small planets like the Earth are difficult to see because they are lost in the glare of the star that they go around. To see Earth-sized planets, we need to be able to put groups of large tele-scopes in orbit around the Earth to get outside the Earth's atmosphere. This will be done in the early part of the 21st century, and soon after we should be able to see whether those Earth-like planets have atmospheres like the Earth's. Astronomers are looking for oxygen in other planet atmospheres, because oxygen is the gas that we breathe. More impor-tantly, all the oxygen on the Earth has come from life, from the plants and the earliest life that appeared on the planet. So, if we find an Earth-sized planet with oxygen in its atmosphere, then it will be a good bet that life is on that planet. This discovery could well be the biggest in the history of astronomy, and it might not be too far off in the future. This is an exciting time.

But, as we said before, life needn't be like that on the Earth. One thing that we have learned about life on the Earth is that it adapts and spreads to every place, no matter how harsh the conditions seem to be. Some creatures live without sunlight in caves. Some live at the bottom of the ocean. Some live in the frozen wastes of the Arctic and the Antarctic. These forms of life are very different from us because they need to live in very different places.

If life can be this different on our planet alone, what kinds of strange life could exist on all the billions of planets that must be circling the stars throughout our Universe?

This is a microscope photograph of a small part of a meteorite. The meteorite, called ALH84001, is thought to have come from Mars but was found in Antarctica on the Earth. Some scientists think that the small 'tubes' in this picture are fossils of tiny life-forms, other scientists aren't convinced.

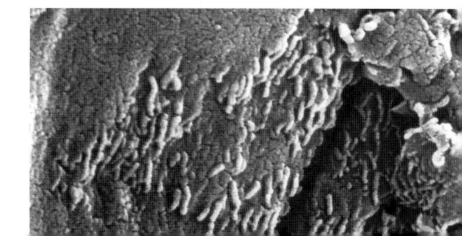

Questions and Answers

The questions in this section were provided by pupils whom the authors of this book have taught. The answers were very kindly supplied by Professor John Brown, Astronomer Royal for Scotland. He has always enjoyed talking to children about space, but how did he become interested and end up doing astronomy as a job? Here is the story in his own words.

My first dim recollection of getting keen on the stars is when I was about 8 when I read a science fiction story which I think was by Patrick Moore. Some folk notice the sky very young. My neighbours' son Dominic Ramsay is about 20 months old but has been pointing at the moon for ages. Recently he pointed at the Space Shuttle and International Space Station and said 'moving, moving' as they whizzed across the sky together!

The first I clearly recall was when I was 10 – in 1957. That year was when Sputnik was launched, when my Uncle Joe showed me Comet Arend Roland through binoculars (it was quite like Comet Hale Bopp in 1997), and the start of Patrick Moore's monthly Sky at Night – the longest running TV programme ever.

John Brown aged about 10.

Around 11 or 12 I got in tow with my dad's photographer friend Eddie Cotogno. He was a bit crazy but filled me full of enthusiasm by telling me stories about the sky (not all true!) showing me Saturn through a small but nice telescope on his Council Flat balcony, and giving me lenses to make telescopes. The first I made were really just a spectacle lens and a magnifying glass taped on opposite ends of two cardboard tubes (from calendars, toilet rolls, etc.) which slid in and out to focus. Later, I built more substantial telescopes, helped and supplied with metalwork bits by my engineering dad. As a family we were never rich, at least not with money, but we had unlimited supplies of support for each other and enthusiasm for doing things. So the fact that telescopes were much more expensive then than now did not get in my way – to see the moon's craters through simple telescopes you make yourself is fantastic!

By the time I was 16 or so I had started a school astronomy club and we later built a telescope for it from a kit. I also started going by bus to a nearby Museum which has an Observatory (Coats, Paisley near Glasgow in Scotland) and attending evening astronomy talks there. Much of this was made possible by the advice and encouragement of my Dumbarton Academy science teachers – Harry ('Cuddy') Mair of Chemistry and especially John McIntyre of Physics. Not only did John get me going to lectures and chasing the headmaster for telescope funds, but it was he who suddenly made physics seem really clear, simple, and related closely to the everyday world. Without that I might not have gone into astronomy professionally.

Around the time I became a student at Glasgow University, I also made a couple of complete telescopes including the mirrors – great fun but taking a lot of time and patience. I studied physics and astronomy (and also maths), physics being the main route these days into astronomy though not the only one. The Astronomer Royal for Scotland before me – Professor Malcolm Longair – studied electronic engineering first. Despite having a lot of studying to do, I managed to keep making telescopes and having fun looking at the sky as an amateur. I also had summer jobs in astronomy – in Edinburgh and Harvard in the USA.

Back then, as now, permanent jobs in astronomy were not that easy to come by and I was fortunate that the then Glasgow Astronomy Professor,

Peter Sweet, seemed to think I was talented. He offered me a teaching job while I was doing my PhD research on the theory of the newly discovered X-rays from the sun. Since then I have had a series of academic posts in Glasgow University and temporary visiting research jobs in many countries, working on a variety of astronomy subjects, not just the sun. In 1984 I was appointed to the new Glasgow Chair of Astrophysics and was by then very active in promoting astronomy by public and school lectures. Then in February 1996 the Queen honoured me by making me *The Astronomer Royal for Scotland*.

John Brown aged about 50.

So now, being the 10th Astronomer Royal for Scotland, besides my regular teaching, research, and management job as a professor at the University of Glasgow, I put a lot of my time into giving talks and planetarium shows in schools and across the country. I also manage to work astronomy into my magic shows and put some into a comedy show at the Edinburgh Fringe Festival. All these activities are a great pleasure for me – above all talking about space to school children. It is a great joy to share with them the pleasure I have been getting from the stars ever since I was their age. I very much hope that readers of this lovely book by Rosie Coleman and Andrew Conway (one of about 25 PhD students I have helped on their way as astronomers, and who has given me lots of help and support) will have as much fun with astronomy as I have had, just for the love of it, and that some may become professionals – maybe Astronomer Royal!

How long does it take to go to a star?
From Colin

The time it takes depends on the distance to the star and how fast you travel. The fastest anything can travel is the speed of light (Einstein told us) which is the huge speed of 300 000 kilometres every second and this is used to describe distances to stars in Light Years – the distance travelled by light in one year – roughly 10 000 000 000 000 (ten million million) kilometres. The distances to stars are measured in various ways. For nearby stars, as the Earth goes round the Sun we can see these stars move slightly compared to fainter, distant ones – just like nearby things move when you move your head (or when you shut one eye then the other). The nearer the star, the bigger the shift. For stars further away this shift is too small to see and we measure their distance by how bright they look. The colour of a star can tell us how bright it really is and how faint it looks then tells us how far away it is. Then knowing the distance we can figure out how long it would take to get there at any chosen speed, just like knowing how long it takes to get from Glasgow to London (400 miles) driving at 50 miles an hour, namely 8 hours.

When is the Sun going to blow up?
From Megan

In a few billion (1 000 000 000) years, so we are safe from that for quite a while. What will happen is that the nuclear reactions in the middle of the Sun which keep it glowing steadily will eventually use up all available fuel (hydrogen). When that happens the reaction stops and the centre of the Sun will shrink while the outer parts expand into a much bigger but cooler star called a red giant which will engulf and vaporise the Earth. So the Sun won't blow up like a bomb – more like a balloon inflating.

What is a super-cluster?
From Martin

Our Sun is one of 100 billion stars making up the Milky Way galaxy which belongs to the local cluster of galaxies including the two Magellanic Clouds, only visible from Australia and other southern places, and the Andromeda Galaxy – the furthest object we can see with our eyes. Well beyond our local cluster are other clusters of galaxies which are clustered around our cluster and the whole lot makes up a super-cluster of galaxy clusters – the Universe is full of such super-clusters. So, if you think of stars as people, then galaxies are cities, clusters are countries and super-clusters are continents that make up the world (or Universe).

What triggers the nuclear explosions when a star is born?
From Ryan

When matter falls under the effect of its weight, it loses what is called gravitational energy and gains speed or kinetic energy. This can be turned into heat: that's how hydroelectric power works – by letting water fall downhill through pipes with turbines in them. Stars are so heavy they fall inward under their *own* weight and heat up. As they fall in, the temperature becomes so great that it triggers nuclear explosions which keep the star hot – and the star is so heavy the explosion energy is contained and only leaks out slowly as sunlight.

How do you become an astronaut?
From Fiona

There are two kinds of astronaut – those who fly spaceships and those who travel in them to work and do experiments. The usual route to become a space pilot is to train as a plane pilot, eventually of high-speed military planes, then move on to special training for space conditions and rocket flight. Those who go into space but are not pilots need only be trained as scientists and get some special fitness and technical training to cope with the special conditions of rocket flight – high acceleration, weightlessness and so on – and

A space shuttle astronaut at work on the international space station.

to work all the complex equipment used in their experiments in space. Some of these experiments are in astronomy but many are to do with engineering, biology, medicine and studying the Earth from space – looking at weather, forest fires, pollution, etc.

Do people think there is a tenth planet?
From Jennifer

Some people do and some don't! It also depends on what you mean by 'planet' and especially whether you include 'minor planets' or asteroids – rocky bodies smaller than Mars or Mercury or Pluto, of which there are already many hundreds known. What most people mean by the tenth planet or 'Planet X' is a large body like the main nine known planets. For many years people thought there was evidence for Planet X in the disturbed movements of the other nine, but this is no longer such a prevalent idea. On the other hand no one would rule out the idea of the existence of Planet X and discovering it remains an exciting possibility for future astronomers – like you!

Will there ever be any kind of animals from Earth going into space?
From Mark

There already have been other animals in space. When the Russians took the lead in the space race, before they launched a human they first launched an unmanned spacecraft, Sputnik 1, in October 1957 then Sputnik 2 in November 1957 with a dog called Laika aboard (Laika means luck in Russian!). Laika was subsequently killed by lethal injection since the space probe could not be returned to Earth. The Americans in turn orbited two monkeys. Since then the treatment of animals in space has been much more humane and will no doubt be so in future.

Is there an end to the Universe or does it go on forever?
From Nicola

We are not sure of the answer to that – it also depends a bit on what you mean by the question! One thing that is for sure is that we cannot see all of the Universe. This is because the whole Universe is expanding. The most distant stars are moving away from us so fast that their light waves get stretched out until it becomes too faint to see. This is the edge of the visible Universe. Another point is that the Universe may be what is called 'closed' but have no edge or end. This is like the surface of the Earth – it does not go on forever but if you travelled on and on across it you would never come to an edge or end – just go around the globe again and again.

How long can you stay in space for?
From Paul

Without an oxygen supply, you would suffocate in a couple of minutes. Without a space suit, you would be exposed to damaging ultraviolet radiation and particles from the sun – you would also become very cold and the low 'air' pressure around you could be damaging in the same way as deep sea divers suffer from 'the bends'. The Hollywood movie idea that you would explode is not true, as is clear from people making high-altitude balloon flights.

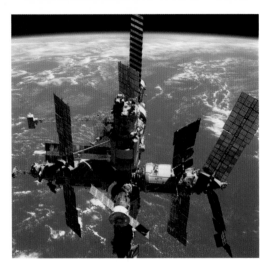

The Mir space station.

In a space suit or a spaceship, you are safe for much longer, so long as your air, food, and water are replenished. The longest stay in space so far was aboard the Russian Mir Space Station by Cosmonaut Valery Polyakov who spent 438 days, or a year and a bit, in space without a break, returning to Earth in March 1995. Food and other supplies were regularly brought from Earth via visiting spacecraft. The only problems with that are that long periods of weightlessness weaken your bones and affect your health. You may also be exposed to dangerous radiation from the Sun. A flight to Mars would take even longer than that and would need to carry supplies for the long trip, so that still needs a lot of planning.

Is it true that from other galaxies our Sun will look like a star?
From Ian

Yes indeed – and a very faint one which you could only see with a very large telescope.

How many stellar systems are there?
From Genna

We don't know. Until about ten years ago we only knew of our own stellar system – a system with a star and planets around it – but we thought there would be others. Recent technology has let us detect (indirectly) the presence of planets round some other stars and the total so far is about 50 and steadily rising. But the fact that we can so far only detect big planets of stars close by suggests that there are many of them out there. Even supposing only one star in every 100 had planets, that would mean about 100 000 000 000 000 000 000 stellar systems in the visible Universe.

In this image of a nebula it is believed that all stages of stellar system formation are taking place. The pillars of glowing gas at the right are newborn stars. If you look carefully you can see dark blobs at the top right corner – these are gas clouds that might collapse to form stars. At the picture's centre lies a cluster of bright hot blue stars that will soon use up their nuclear fuel. A bright supergiant star, above and to the left of the cluster, is coming to the end of its life. It is surrounded by a glowing ring and may have only a few thousand years left before it goes supernova. Below the star cluster and to the right of the main bit of the nebula you can see orange blobs which, like the proplyds in Orion, may be the beginning of a new stellar system: a star together with planets.

How did the Solar System form?
From Kimberley

The Sun formed by the shrinking of a big gas cloud under its own weight. As it shrank it heated up and started to shine (a bit like hydroelectric power in which falling water generates electricity) but it also spun faster like a skater pulling in her arms. The spin become so fast it threw the outer layers of the Sun off (like a playground roundabout pushes you out) and they formed a disk of gas and dust, which cooled and formed the planets, comets and meteors.

How did the dust form to make the planets?
From Alex

A good question and the answer is that we are not very sure. There is lots of dust all over the Universe, even in quite hot places where we might expect it to be vapourised and how it survives is not well-known. Roughly speaking, dust forms when hot gas cools down and turns to liquid then to solid – like water vapour in the air turns to raindrops then hailstones or snowflakes when the vapour is suddenly cooled (like water drops form when you (carefully) hold a cold plate above a steaming kettle). When the Sun was forming from a gas cloud it shrank due to its own gravity (weight) and started to spin faster and faster like a skater. This spun gas off the Sun which flew out and cooled down, turning into dust. There was so much of this dust that bits were pulled together by gravity and formed bigger and bigger chunks forming meteors, asteroids, planets and (ice/dust) comets.

If the Sun is a star, how does it come out in the daytime and other stars come out at night?
From Lisa

The Sun does not really 'come out'. What happens is that when the Sun rises (as the Earth turns) it is so bright that when its light gets scattered in the air we breathe it makes the sky glow bright blue (or white/grey if there are clouds). This glow hides the stars in the daytime though they are actually there (you can see the bright ones in a telescope in the daytime). If there was no atmosphere the sky would not be bright in the daytime, but black like at night and you would see the stars in the daytime (as astronauts in space and on the Moon do). Likewise, if the other stars were much brighter they would make the sky glow and tend to hide the fainter stars – just like the Full Moon does. It's all because the Sun is so very, very bright and because Earth has an atmosphere.

How do you know there's a black hole?
From Kerry

Black holes cannot be seen directly since they do not emit light and absorb all light that falls on them. But their presence can be detected indirectly by other means, including these.

1. Bending the light of objects behind them and giving us distorted images of distant galaxies.
2. Pulling matter in, which heats up as it falls and becomes so hot it glows in X-rays which we can see.
3. The speed of stars or gas around a black hole is very high because the black hole is heavy. This lets us 'weigh' the object and we find it is so small and heavy that the only thing it can be is a black hole.

Can you explain the activity inside a black hole?
From Martin

We don't, and can't, see activity actually inside a black hole since no light can get out of it. If we went inside to see it, we would get crushed by the huge gravity. But theory says that things inside a black hole would be squashed up and moving very fast so in that sense we know it is active. What we have seen is 'activity' of matter just outside a black hole as it falls in.

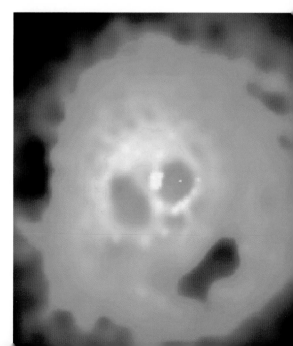

This is an image of X-rays coming from a cluster of galaxies. The brightest part at the centre is probably the location of a massive black hole.

Have we got any evidence that there are aliens on Mars?
From Mark

No, none at all. Certainly not 'aliens' in the sense of strange animals. There is some slight fossil evidence that there may have been and might still be microscopic life forms on Mars, but the evidence is not yet clear at all.

Is the Moon burning gas like the Sun and stars?
From Nicole

No – the Moon is much colder than the Sun or stars and so emits very little light of its own. The moonlight we see is light from the Sun reflected off the Moon's surface – just as is the light we see from the ground and the clouds on Earth.

How big is a crater on the Moon?
From Shaun

They range from tiny, tiny pits a few millimetres in size up to very large sizes. The biggest diameter crater on the visible side of the Moon is Bailly which is 183 miles (295 km) across with walls 14 000 feet (4250 m) high. The Orientale Basin, part of which is on our side and part on the hidden side of the Moon is 600 miles (965 km) across. The deepest crater is Newton, the walls of which rise to 29 000 feet (8850 metres) above its central floor.

Craters on the Moon.

How far is the Moon from Earth?
From Daniel

About quarter of a million (250 000) miles or 400 000 km, though it varies slightly since the Moon's orbit is not quite a circle. The distance is actually increasing but very, very slowly. Though that sounds far it is not very far compared to other astronomical objects – even the Sun is 400 000 times further and the nearest star is more than 200 000 times further than that. A train on the Glasgow–London route may make two return trips a day and that is about 1600 miles, so in a year a Glasgow–London train travels a distance equal to that from Earth to Moon and back.

Will there be football stadiums on the Moon?
From Jason

Maybe eventually, once there are big enough bases set up to have air to breathe over the large area of a football field. It will of course have to be indoor football and tickets are likely to be very, very expensive. We might need to change the size and the rules since gravity is so low that it would be easy even for a child to kick a ball from one goal to the other.

How long does it take to get to the Moon?
From Jamie

That depends how fast you travel. At the speed of a fast train (if there were rails) it would take a few months. At the huge speed of light – 300 000 km per second – it takes about a second. For a typical spacecraft it is in between these numbers and takes around two days. At walking speed it would take ten years.

How could you take photographs of Venus when it is so hot?
From Debra (who wants to be an astronomer)

I assume you mean on its surface – the clouds are so thick you have to land there to see the surface. That is difficult but has been done by the Russian Venera spacecraft. You have to make a spacecraft and camera of strong heat-resistant metals and glass so that they don't melt in the heat or get crushed by the pressure before the pictures are taken and radioed back.

If instead you mean just to take a picture of the whole planet, you can do that from here using a telescope and ordinary camera or closer up from a space probe orbiting or flying by Venus. In both these cases you are above the clouds of Venus, well away from the solid surface so the heat is not a problem – just like photographing a hot light bulb from a distance.

Why did you go to space in the first place?
From Thomas

I personally have never 'gone to space' though I would love to for the thrill and the view back to Earth. Folk who have gone there did it for those reasons and to carry out experiments you can't do on Earth because of the air and gravity.

Is it true that if we went close enough to Jupiter it would suck us in like a vacuum cleaner?
From Siobhan

Definitely not. If you were close in and not moving sideways to orbit the planet, you would fall in under gravity, just as you do on Earth, but quickly because Jupiter's gravity is strong. But there is no 'sucking' action!

Is it true that aliens landed in Roswell?
From Christopher

Personally I don't believe a word of it and think people who do are fooling themselves.

What stops neutron stars from getting any smaller than they end up and becoming black holes?

From Daniella

A very good question. It's really because matter can exist in different forms called states. The states are *solid*, like a brick, *liquid*, like water, and *gas*, like air.

If you took a huge balloon full of gas (one state) and you squeezed it down smaller and smaller, it would eventually turn into a liquid and you could not really compress it any further. You can compress gases, but not solid or liquids.

A normal star is made of gas. If you take a big star and squeeze it (with the help of gravity)

The star at the end of the arrow is a neutron star. It might not look much, even to the Hubble space telescope, but it is in fact more massive than our Sun but only about 20 km across.

down from being ten million km across down to around ten km, it will no longer be made of gas. It will become a neutron star – an incredibly dense solid made of tiny particles called neutrons. So, by being a solid, and a rather unusually strong solid at that, the star can resist the pull of gravity and not shrink any further.

Not all stars will end up as neutron stars. Many stars, like the Sun, have too little mass and so don't have enough gravity to pull them down to the size of a neutron star. They end their lives as White Dwarves (that's another story). More massive stars have so much gravity that they are pulled down to a size smaller than a neutron star – they will become black holes.

What can't astronomers explain?
From Stephen

Lots and lots of things! In fact astronomers and scientists in general don't really ever *explain* anything, they just find simple ways to *describe* things we see that have something in common.

For example, Newton realised that apples fall down from trees and the Moon falls toward the earth because they are pulled by the same thing which he called gravity. His *Law of Gravity* told us how to describe how everything from apples and rockets to stars and galaxies will move – but it does not explain *why* there is gravity. The famous physicist Einstein came up with a better theory of gravity, called *relativity*, but even that doesn't answer the *why* question completely. In fact, it may be that questions like that do not have a final answer. You could say 'God made gravity that way' but then someone might ask you why he chose that rather than something different.

Perhaps you are really asking what astronomers can't describe easily *yet*. The answer again is a lot of things, though there is much we do understand. Here are some big questions that I think still need answers.

1. How do we account for the very odd rotation of some planets – notably Venus (very slow and backwards) and Uranus (fast but tilted at right angles)?
2. How do flares on the sun work?
3. How do some planets manage to exist near some pulsars when we know that pulsars came from supernova explosions?
4. How exactly do stars form?
5. Are there black holes in the middle of all galaxies and, if so, why?
6. Is the universe open or closed – will it expand forever or not?

Index

aliens   *see life*
Andromeda Galaxy   5, 25, 123
asteroid belt   82
astrology   14
astronauts   78, 134
astronomy   2, 14
aurora   58
axis   71

big bang   61, 126–128
big crunch   128
black holes   113, 139, 143, 144
brown dwarfs   107

calendar of universe   61
circumpolar   7
comets   91, 93–97, 130
constellation   2, 10
Constellations
    Andromeda   19, 25
    Big Dipper   11
    Canis Major   34
    Canis Minor   35
    Carina   39
    Cassiopeia   19
    Centaurus   12, 43
    Crux   41
    Cygnus   10, 27
    Lyra   27
    naming   11
    Orion   10, 33
    Pegasus   25
    Perseus   23
    Plough   11, 17

    Scorpius   10, 37
    Southern Cross   *see Crux*
    summer   10
    Taurus   10, 31, 33
    Ursa Major   11
    winter   10
craters   65, 74, 92, 140

day   71, 138

eclipse
    lunar   78
    solar   60
ellipse   62
Equator   6–7
extra-terrestrials   *see life*

finite   126
fireball   91

galaxies   121
gravity   50
greenhouse effect   66

Hubble Space Telescope   13, 108

infinite   126

Leap Year   72
life   128–129, 140, 142
light-year   4
local group   125

Magellanic clouds   38–39
magnitude   12

mass   50
meteorite   44, 91–92, 129
meteoroids   93
meteors   91
Milky Way   19, 122–124
Moon
    astronauts   78
    distance to   141
    eclipse   78
    football   141
    lunar month   76
    phases   76–77
moons   64

nebula
    Crab   108, 119
    Horsehead   118
    Orion   33, 115
    Pleiades   115
    proplyds   117
    birth of stars   52
neutron star   110
night   71, 138
north star   9
Northern Hemisphere   6–7, 72
northern lights   *see aurora*
nova   114
nuclear fusion   55, 107

orbit   62

pangea   70
photosphere   55
planets
    Earth   69–73, 113
    Jovian   62
    Jupiter   83–84, 142
    Mars   80–81
    Mercury   64–65
    Neptune   88–89
    Planet X   135
    Pluto   90–91
    Saturn   85–86
    Terrestrial   62
    Uranus   86–87, 144
    Venus   66–68, 142, 144
planisphere   14
Pole Star   9
pulsars   112, 144

red giant   98, 110

satellite   44, 73
seasons   10–11, 72
shooting star   *see meteorite*
Solar System   1, 46
Southern Hemisphere   6–7, 72
southern lights   *see aurora*
space shuttle   44, 59, 130
spectrum   102–103
speed of light   4, 132
star clusters
    globular   29, 120–121
    Hyades   31
    M13   29, 120
    open   120
    Perseus   23
    Pleiades   31, 115, 120
    Seven sisters   *see Pleiades*
stars
    Aldeberan   12, 31
    Algol   23
    Alpha Centauri   4, 12, 43
    Altair   27
    Antares   37
    Betelgeuse   33, 101
    birth   51, 106, 144
    colours   101
    death   98, 106, 108, 111, 133
    Deneb   27
    moving   44–45
    names   12
    Procyon   35
    Proxima Centauri   43, 113
    Rigel   34, 101, 111
    Rigil Kent   *see Alpha Centauri*
    Sirius   13, 34, 111, 113
    Sun   54, 111, 113, 144
    variable   23
    Vega   27
star trails   8–9
Summer Triangle   27
sunspots   55–56
super-clusters   125, 133
supernova   108–110

Universe   126, 135, 144

Voyager   63

white dwarf   35, 110

zodiac   13

Acknowledgements

2, Linda Davison; 3, 16, 18, 22, 24, 26, 28, 30, 32, 33*t*, 36, 40, 42, Till Credner and Sven Kohle, AlltheSky.com; 38*t*, 38*b*, Sven Kohle, AlltheSky.com; 6, 11, 17, 19, 20, 21, 23, 25, 27, 29*b*, 31*b*, 33*b*, 34, 35, 37, 39, 41, 43, 47, 49, 53, 60, 61, 70, 71, 77, 78, 111, 113, 123, Mark Garlick; 8, 109, © Anglo-Australian Observatory – photography by David Malin; 13, 48, 59, 80, 81, 105, 106, 110, 129, 134, 136, NASA; 29*t*, 120, Y. Kitahara; 31*t*, © Anglo-Australian Observatory/Royal Observatory, Edinburgh; 51, C. O'Dell and S. Wong, Hubble Space Telescope, NASA; 56*t*, Ray Gralak; 56*b*, P. Brandt, G. Scharmer, R. Shine, G.W. Simon; 57, HAO eclipse team; 58, Wade B. Clark Jr.; 63, 88, JPL, NASA; 65, M. Robinson, Mariner 10, NASA; 67, Galileo Project, NASA; 68, E. de Jong et al, Magellan, NASA; 73, 74, 75, Apollo 17, NASA; 79, Apollo 11, NASA; 82, JHU APL, NASA; 83, Galileo, JPL, NASA; 84, R. Beeke, A. Simon, NASA; 85, Hubble Heritage, NASA; 87, H. Hammel, MIT, NASA; 89, Voyager 2, NASA; 90, R. Albrecht, NASA; 92, D. Noddy, USGS; 94, The Astronomy Class at EUC-Syd and Amtsgymnasiet in Sonderberg, Denmark – www.amstgym-sdbg.dk/as; 95, H.A. Weaver, T.E. Smith, Hubble Space Telescope, NASA; 97, Don Davis, NASA; 100, ESA/NASA; 102, Damon Hart-Davis; 103, S. Best, Open University; 115, David Malin; 116, Bill Schoening/NOAD/AURA/NJF; 117, C.R. O'Dell, NASA; 118, 121, 125*b*, N.A. Sharp/NOAD/AURA/NJF; 119, W.P. Blair, JHU/NASA/ AURA; 124*t*, B. Schoening, V. Harvey/REU program/NOAD/AURA/NJF; 124*b*, Todd Boroson/NOAD/AURA/NJF; 125*t*, NOAD/AURA/NJF; 127, R. Williams/ HDF team/NASA; 137, W. Brandner, E. Grebel, Y-H Chu, NASA; 139, A. Fabian et al, NASA; 140, E. de Jong, Magellan, NASA; 143, F. Walter, NASA